T0141919

Advances in Intelligent Systems and Computing

Volume 1005

The series "Advances in Intelligent Systems and Computing" contains publications on theory, applications, and design methods of Intelligent Systems and Intelligent Computing. Virtually all disciplines such as engineering, natural sciences, computer and information science, ICT, economics, business, e-commerce, environment, healthcare, life science are covered. The list of topics spans all the areas of modern intelligent systems and computing such as: computational intelligence, soft computing including neural networks, fuzzy systems, evolutionary computing and the fusion of these paradigms, social intelligence, ambient intelligence, computational neuroscience, artificial life, virtual worlds and society, cognitive science and systems, Perception and Vision, DNA and immune based systems, self-organizing and adaptive systems, e-Learning and teaching, human-centered and human-centric computing, recommender systems, intelligent control, robotics and mechatronics including human-machine teaming, knowledge-based paradigms, learning paradigms, machine ethics, intelligent data analysis, knowledge management, intelligent agents, intelligent decision making and support, intelligent network security, trust management, interactive entertainment, Web intelligence and multimedia.

The publications within "Advances in Intelligent Systems and Computing" are primarily proceedings of important conferences, symposia and congresses. They cover significant recent developments in the field, both of a foundational and applicable character. An important characteristic feature of the series is the short publication time and world-wide distribution. This permits a rapid and broad dissemination of research results.

**** Indexing: The books of this series are submitted to ISI Proceedings, EI-Compendex, DBLP, SCOPUS, Google Scholar and Springerlink ****

More information about this series at http://www.springer.com/series/11156

Florentino Fdez-Riverola ·
Miguel Rocha · Mohd Saberi Mohamad ·
Nazar Zaki · José A. Castellanos-Garzón
Editors

Practical Applications of Computational Biology and Bioinformatics, 13th International Conference

Editors
Florentino Fdez-Riverola
Edificio Politécnico
Escuela Superior de Ingeniería Informática,
Campus Universitario As Lagoas
Ourense, Spain

Mohd Saberi Mohamad
Faculty of Bioengineering and Technology
Universiti Malaysia Kelantan
Kelantan, Malaysia

José A. Castellanos-Garzón
IBSAL/BISITE Research Group
University of Salamanca
Salamanca, Salamanca, Spain

Miguel Rocha
Department de Informática
Universidade do Minho
Braga, Portugal

Nazar Zaki
Department of Computer Science
and Software Engineering Leader,
Data Science Research Group
College of Information Technology (CIT)
United Arab Emirates University (UAEU)
Al Ain, United Arab Emirates

ISSN 2194-5357 ISSN 2194-5365 (electronic)
Advances in Intelligent Systems and Computing
ISBN 978-3-030-23872-8 ISBN 978-3-030-23873-5 (eBook)
https://doi.org/10.1007/978-3-030-23873-5

This Springer imprint is published by the registered company Springer Nature Switzerland AG
The registered company address is: Gewerbestrasse 11, 6330 Cham, Switzerland

Preface

Next-generation sequencing technologies, together with other emerging and quite diverse experimental techniques, are evolving rapidly, creating numerous types of omics data. These are creating new challenges for the expanding fields of bioinformatics and computational biology, which seek to analyze, process, integrate, and extract meaningful knowledge from these data. This calls for new algorithms and approaches from fields such as databases, statistics, data mining, machine learning, optimization, computer science, machine learning, and artificial intelligence. Clearly, biology is increasingly becoming a science of information, requiring tools from the computational sciences. To address these challenges, we have seen the surge of a new generation of interdisciplinary scientists with a strong background in the biological and computational sciences.

The International Conference on Practical Applications of Computational Biology & Bioinformatics (PACBB) is an annual international meeting dedicated to emerging and challenging applied research in bioinformatics and computational biology. Building on the success of previous events, this volume gathers the accepted contributions for the 13th edition of the PACBB Conference after being reviewed by different reviewers, from an international committee from 21 countries. PACBB'19 technical program includes 21 papers from authors of many countries (Australia, Colombia, Egypt, Germany, India, Malaysia, Portugal, Saudi Arabia, Slovakia, South Korea, Spain, Switzerland, Turkey, United Arab Emirates, UK, and USA) and different subfields in bioinformatics and computational biology. There will be special issues in JCR-ranked journals such as Interdisciplinary Sciences: Computational Life Sciences, Integrative Bioinformatics, Information Fusion, Neurocomputing, Sensors, Processes, and Electronics. Therefore, this event will strongly promote the interaction of researchers from diverse fields and distinct international research groups. The scientific content will be challenging and will promote the improvement of the valuable work that is being carried out by the participants.

This symposium was organized by the University of Malaysia Kelantan, University of Minho, University of Vigo, and University of Salamanca. This edition was held in Avila, Spain, from June 26–28, 2019. We thank the sponsors (IEEE

Systems Man and Cybernetics Society Spain Section Chapter and the IEEE Spain Section (Technical Co-Sponsor), IBM, Indra, Viewnext, Global exchange, AEPIA, APPI and AIR Institute), the funding supporting of the with the project *"Intelligent and sustainable mobility supported by multi-agent systems and edge computing"* (Id. RTI2018-095390-B-C32), and finally, the Local Organization members and the Program Committee members for their hard work, which was essential for the success of PACBB'19.

Florentino Fdez-Riverola
Miguel Rocha
Mohd Saberi Mohamad
Nazar Zaki
José A. Castellanos-Garzón

Organization

General Co-chairs

Mohd Saberi Mohamad	Universiti Malaysia Kelantan, Malaysia
Miguel Rocha	University of Minho, Portugal
Florentino Fdez-Riverola	University of Vigo, Spain
Nazar Zaki	United Arab Emirates University, United Arab Emirates
José Antonio Castellanos Garzón	University of Salamanca, Spain

Program Committee

Vera Afreixo	University of Aveiro, Portugal
Amparo Alonso-Betanzos	University of A Coruña, Spain
Rene Alquezar	Technical University of Catalonia, Spain
Manuel Álvarez Díaz	University of A Coruña, Spain
Jeferson Arango Lopez	Universidad de Caldas, Colombia
Joel Arrais	University of Coimbra, Portugal
Julio Banga	Instituto de Investigaciones Marinas (CSIC), Spain
Carlos Bastos	University of Aveiro, Portugal
Carole Bernon	IRIT/UPS, France
Lourdes Borrajo	University of Vigo, Spain
Ana Cristina Braga	University of Minho, Portugal
Boris Brimkov	Rice University, USA
Guillermo Calderon	Autonomous University of Manizales, Colombia
Rui Camacho	University of Porto, Portugal
José Antonio Castellanos Garzón	University of Salamanca, Spain

Luis Fernando Castillo	Universidad de Caldas, Colombia
José Manuel Colom	University of Zaragoza, Spain
Fernanda Brito Correia	DETI/IEETA, University of Aveiro, and DEIS/ISEC/Polytechnic Institute of Coimbra, Portugal
Daniela Correia	University of Minho, Portugal
Roberto Costumero	Technical University of Madrid, Spain
Francisco Couto	Faculty of Sciences, University of Lisbon, Portugal
Yingbo Cui	National University of Defense Technology, China
Masoud Daneshtalab	KTH Royal Institute of Technology in Stockholm, Sweden
Javier De Las Rivas	University of Salamanca, Spain
Sergio Deusdado	Polytechnic Institute of Bragança, Portugal
Oscar Dias	University of Minho, Portugal
Fernando Diaz	University of Valladolid, Spain
Ramón Doallo	University of A Coruña, Spain
Xavier Domingo-Almenara	Rovira i Virgili University, Spain
Pedro Ferreira	Ipatimup—Institute of Molecular Pathology and Immunology of the University of Porto, Portugal
João Diogo Ferreira	Faculty of Sciences, University of Lisbon, Portugal
Nuno Filipe	University of Porto, Portugal
Mohd Firdaus-Raih	National University of Malaysia, Malaysia
Nuno A. Fonseca	University of Porto, Portugal
Dino Franklin	Federal University of Uberlandia, Spain
Alvaro Gaitan	Café de Colombia, Colombia
Narmer Galeano	Universidad Catolica de Manizales, Colombia
Vanessa Maria Gervin	Hathor Group, Brazil
Rosalba Giugno	University of Verona, Italy
Josep Gómez	Rovira i Virgili University, Spain
Patricia Gonzalez	University of A Coruña, Spain
Consuelo Gonzalo-Martin	Universidad Politécnica de Madrid, Spain
David Hoksza	Univerzita Karlov, Czech Republic
Natthakan Iam-On	Mae Fah Luang University, Thailand
Gustavo Isaza	University of Caldas, Colombia
Paula Jorge	University of Minho, Portugal
Martin Krallinger	National Center for Oncological Research, Spain
Rosalia Laza	Universidade de Vigo, Spain
Thierry Lecroq	University of Rouen, France
Giovani Librelotto	Federal University of Santa Maria, Portugal
Filipe Liu	CEB, University of Minho, Portugal
Ruben Lopez-Cortes	University of Vigo, Spain

Eduardo Valente	IPCB, Portugal
Alfredo Vellido	Technical University of Catalonia, Spain
Jorge Vieira	University of Porto, Portugal
Alejandro F. Villaverde	Instituto de Investigaciones Marinas (CSIC), Spain
Pierpaolo Vittorini	University of L'Aquila—Department of Life, Health, and Environmental Sciences, Italy

Organizing Committee

Juan Manuel Corchado Rodríguez	University of Salamanca, Spain and AIR institute, Spain
José Antonio Castellanos	University of Salamanca, Spain
Sara Rodríguez González	University of Salamanca, Spain
Fernando De la Prieta	University of Salamanca, Spain
Sonsoles Pérez Gómez	University of Salamanca, Spain
Benjamín Arias Pérez	University of Salamanca, Spain
Javier Prieto Tejedor	University of Salamanca, Spain and AIR institute, Spain
Pablo Chamoso Santos	University of Salamanca, Spain
Amin Shokri Gazafroudi	University of Salamanca, Spain
Alfonso González Briones	University of Salamanca, Spain and AIR institute, Spain
Yeray Mezquita Martín	University of Salamanca, Spain
Enrique Goyenechea	University of Salamanca, Spain
Javier J. Martín Limorti	University of Salamanca, Spain
Alberto Rivas Camacho	University of Salamanca, Spain
Ines Sitton Candanedo	University of Salamanca, Spain
Daniel López Sánchez	University of Salamanca, Spain
Elena Hernández Nieves	University of Salamanca, Spain
Beatriz Bellido	University of Salamanca, Spain
María Alonso	University of Salamanca, Spain
Diego Valdeolmillos	University of Salamanca, Spain and AIR institute, Spain
Roberto Casado Vara	University of Salamanca, Spain
Sergio Marquez	University of Salamanca, Spain
Guillermo Hernández González	University of Salamanca, Spain
Mehmet Ozturk	University of Salamanca, Spain
Luis Carlos Martínez de Iturrate	University of Salamanca, Spain and AIR institute, Spain
Ricardo S. Alonso Rincón	University of Salamanca, Spain
Javier Parra	University of Salamanca, Spain
Niloufar Shoeibi	University of Salamanca, Spain

Zakieh Alizadeh-Sani	University of Salamanca, Spain
Jesús Ángel Román Gallego	University of Salamanca, Spain
Angélica González Arrieta	University of Salamanca, Spain
José Rafael García-Bermejo Giner	University of Salamanca, Spain

Contents

ProtRet: A Web Server for Retrieving Proteins in a Functional Complex

Nazar Zaki[1(✉)], Maryam Al Yammahi[2], and Tetiana Habuza[1]

[1] Department of Computer Science and Software Engineering, College of Information Technology, United Arab Emirates University (UAEU), Al Ain, United Arab Emirates
{nzaki,201890064}@uaeu.ac.ae

[2] Department of Computer and Network Engineering, College of Information Technology, United Arab Emirates University (UAEU), Al Ain, United Arab Emirates
ms-alyammahi@uaeu.ac.ae

Abstract. The identification of protein complexes is becoming increasingly important to our understanding of cellular functionality. However, if a biologist wishes to investigate a certain protein, currently no method exists to assist him/her to accurately retrieve the possible protein partners that are expected to be in the same functional complex. Here, we introduce ProtRet, a web server that functions as an interface for an improved Pigeonhole approach to identify protein complexes in protein-protein interaction networks. The approach provides high-quality protein comparison that is particularly valuable because of its accurate statistical estimates based on fuzzy criterion and Hamming distance. The proposed method was tested on two high-throughput experimental protein-protein interaction data sets and two gold standard data sets and was able to retrieve more correct protein members than all existing methods. The web server is accessible from the link http://www.protret.com/.

Keywords: Protein complexes · Protein-protein interaction · Pigeonhole method · Information retrieval

1 Introduction

The more we know about the molecular biology of the cell, the more we find genes and proteins as part of networks, sub-network or pathways instead of as isolated entities, and their function as a variable dependent of the cellular context and not only of the individual properties [1]. A protein complex is well-known as a group of two or more polypeptide chains that interact with each other. Detecting the protein complexes is important as they play crucial roles in many important biological processes. All cellular functions, such as cell cycle control, differentiation, signaling, protein folding, translation, transcription, post-translational modification, control of gene expression, inhibition of enzymes, antigen-antibody

F. Fdez-Riverola et al. (Eds.): PACBB 2019, AISC 1005, pp. 1–7, 2020.
https://doi.org/10.1007/978-3-030-23873-5_1

interaction and transportation, are achieved through the formation of protein complexes [2,3]. Therefore, the construction or destruction of certain protein complexes can result in initiation, modulation or termination of certain biological processes [4].

Many successful approaches have been proposed in the past to detect the protein complexes existed in the protein-protein interaction (PPI) networks, such as Molecular Complex Detection (MCODE) [5], CFinder [6] which detects overlapping dense groups, ProRank+ [3] as an extension of the ProRank method which quantifies the importance of each protein based on the interaction structure and the evolutionarily relationships between proteins in the network [7], clustering-based on maximal cliques (CMC) [8], clustering with overlapping neighborhood expansion (ClusterONE) [9], PEWCC which based on reliability assessment and weighted clustering coefficient [10], Restricted Neighbourhood Search Clustering Algorithm (RNSC) [11], Markov Clustering (MCL) [12] and repeated random walks on genome-scale protein networks for local cluster discovery (RRW) [13].

Although all of the aforementioned approaches are able to detect protein complexes to an acceptable accuracy, one question still remains: if a biologist is investigating a certain protein, how can they accurately retrieve the possible protein partners expected to be in the same functional complex? None of the approaches listed have been re-engineered to provide this information. To the best of our knowledge, no method em-ploys the notion of information retrieval to detect protein complexes in PPI networks. In this paper, we introduce ProtRet, a web server that uses the Pigeonhole approach to detect protein complexes in PPI networks.

2 Methods

The retrieval of possible members of a protein complex based on fuzzy criterion essentially depends on the representation of the proteins in terms of their PPI network. We first start by converting the PPI network into adjacency matrix A such that its element A_{ij} is equal to 1 when P_i interacts with P_j, and 0 otherwise. Each row of the adjacency matrix A is referred to by the term Bit-Attribute Vector (BAV) and hence the list of all BAVs constitutes a Bit-Attribute Matrix (BAM). Let's assume that we have a query protein (a BAV of length m) and we want to retrieve its closest proteins in the BAM, which consists of n proteins from P_1 to P_n. First, each item is characterized according to its m attributes from A_1 to A_m. Then, the closeness of proteins is determined by a particular Hamming distance (HD). Traditionally, this is done by a tedious sequential search. We propose utilizing fuzzy search methods such as the Pigeonhole Principle [14]. The Pigeonhole Principle states that if n items (proteins) are put into k containers, with $n > k$, then at least one container must contain more than one item. Accordingly, if we have k segments and $(k - 1)$ mismatches, then there must be at least one segment with no mismatches (exact match) or fewer mismatches than the other segments. Therefore, the segments can be efficiently located if they are within an HD equal to 0 from a given search pattern. For each such match in each segment, we perform a sequential search and compare the remaining segments within a specified threshold t of mismatches. Thus, if l is the least number of mismatches that a segment can have, then, in this case, t can be calculated as:

$$t = [k \times (l + 1)] - 1 \qquad (1)$$

As a result, all proteins within this HD can be retrieved. Therefore, the first step is to partition the BAVs into (k) segments, each of $\frac{m}{k}$ in bit length, where m is the number of columns in the matrix. For each segment, a dictionary table (DT) of a vertical length of $2^{\frac{m}{k}}$ is created. The DT contains all possible values of a segment's BAVs as indexes and the list of BAV locations in the BAM. Each segment is accessed through its corresponding DT to find its location. Thus, the union of all the locations of all the segments will be the candidate list, which will be searched sequentially to collect all the vectors that are less than or equal to the threshold t. In Fig. 1 we illustrate how ProtRet method works.

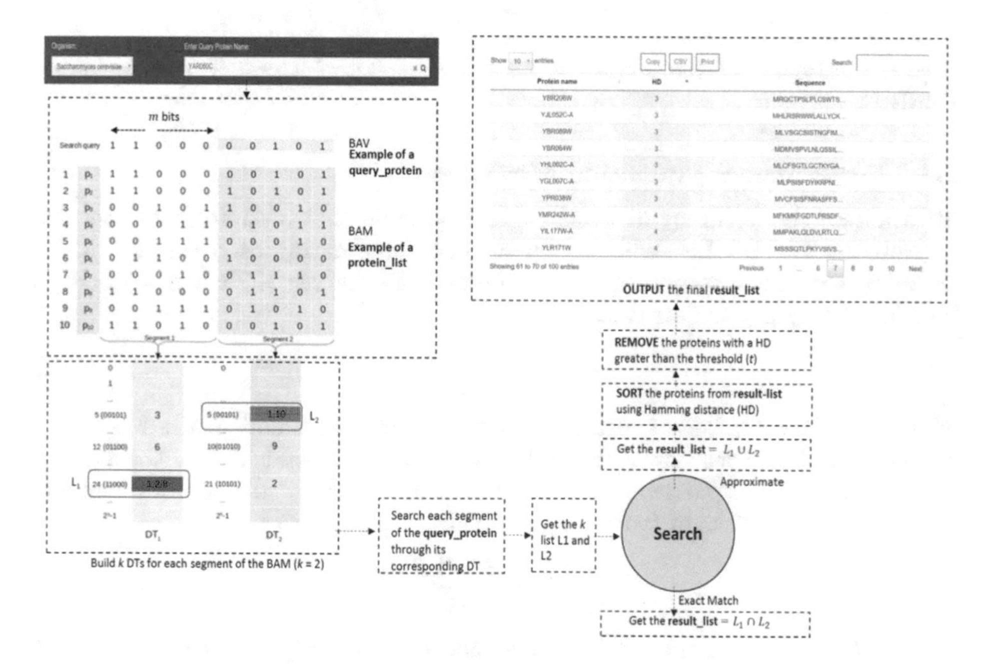

Fig. 1. Overview of ProtRet. The user starts the process by selecting an organism from the drop-down list and then enters a query protein. Auto-complete functionality assists the user to locate the desired query protein more easily and quickly and the result is a list of all possible members of a complex based on their HD.

In addition, the process of finding members of a protein complex by adopting advanced Pigeonhole search method is shown in Fig. 2.

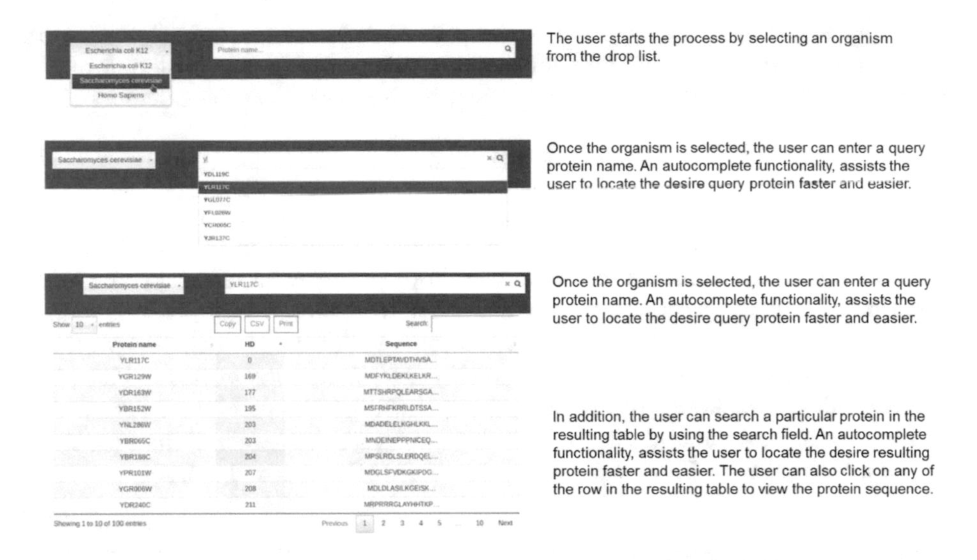

Fig. 2. Illustration of the process of finding members of a protein complex using ProtRet

2.1 Performance Evaluation Measures

To evaluate the accuracy of ProtRet, let's assume that a complex C contains n proteins. The method will run for n times considering each protein in the complex as a query protein. In this example, the accuracy AC when a protein p belonging to complex C is calculated as:

$$AC(p_c) = \frac{x}{n-1} \tag{2}$$

where x is the number of correctly retrieved proteins found in the top $n-1$ proteins list. The accuracy of the protein complex retrieval is simply the average accuracy calculated. To determine the quality of the retrieved complex members, we also used the Jaccard index [10], defined as follows:

$$Match = \frac{|S_C \cap S_R|}{|S_C \cup S_R|} \tag{3}$$

where C is a cluster and R is a reference complex. S_C and S_R are the set of proteins in C and R, respectively. The complex C is defined to match the complex R if $Match \geq \mu$ where $\mu = \{0.1, 0.2, ..., 0.9,\ or\ 1.0\}$, since different methods were evaluated with different values of μ. However, in this study, we will only consider $\mu = 0.25$ since most of the existing methods are evaluated against the same value.

3 Testing the Performance of ProtRet

The performance of ProtRet was tested on two high-throughput experimental PPI data sets (D1 and D2) developed by Gavin in 2002 [15] and 2006 [16],

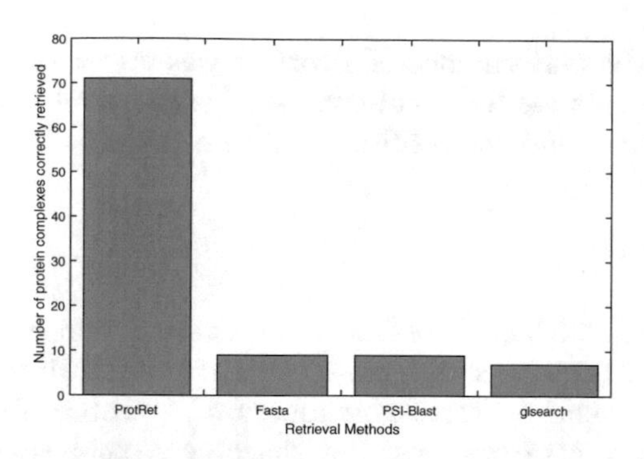

Fig. 3. Comparison of ProtRet to Fasta, PSI-Blast, and glsearch in terms of a number of complexes predicted correctly ($\mu = 0.3$) and considering data set D1 and reference data set Ref1

Table 1. Comparison of the performance of the new method with MCL, MCode, ClusterOne, CFinder, RRW, RNSC and CMC methods, assessed by a number of correctly predicted matched complexes. The value of $\mu = 0.25$ and considering data set D2 and reference data set Ref2

Methods	Number of matched complexes
ProtRet	100
MCL	79
Mcode	65
ClusterOne	82
CFinder	65
RRW	76
RNSC	72
CMC	75

respectively. D1 contains 1,430 proteins and 6,531 interactions and D2 contains 1,855 proteins and 7,669 interactions. In addition, we used Ref1 (81 complexes) [17] and Ref2 (203 complexes) [18] as gold standard reference data sets. For all the data sets used, we set k to 65, each segment length to $\frac{m}{k}$, the dimension of the DT as $2^{\frac{m}{k}}$ and the threshold t as $k - 1 = 64$. The performance of the proposed method was compared with some of the more widely-used sequence-based comparison methods known for their accurate statistical estimates and high-quality alignments, such as FASTA [19], PSI-BLAST [20] and GLSEARCH [21]. Using data set D1, reference data set Ref1 and $\mu = 0.3$ ProtRet was able to retrieve 71 protein complexes correctly compared with 9 retrieved by PSI-BLAST, 9 by FASTA and 7 by GLSEARCH as shown in Fig. 3. As expected, the performance of sequence-based comparison methods was quite poor. This is a good indication that members of a protein complex are not necessarily homologous to each other.

Furthermore, the performance of ProtRet was tested on data set D2 along with the reference data set Ref2 and compared with current methods for detecting protein complexes and the results are shown in Table 1.

4 Conclusion

In this paper, we introduce ProtRet, a web server that uses the Pigeonhole approach to detect protein complexes in PPI networks. ProtRet makes it easy for researchers to identify possible members of a protein functional complex. Currently, only data sets from yeast (Saccharomyces cerevisiae) are included. In the future, we aim to introduce more species' data sets, particularly human data sets.

Acknowledgments. The authors acknowledge financial support from the ICT Fund (Grant # G00001472) and the UAEU (Grant # G00002659).

References

1. Zaki, N., Deris, S., Alashwal, A.: Protein-protein interaction detection based on substring sensitivity measure. Int. J. Biomed. Sci. **1**(2) (2006). ISSN 1306–1216
2. Farutin, V., Robinson, K., Lightcap, E., Dancik, V., Ruttenberg, A., Letovsky, S., Pradines, J.: Edge-count probabilities for the identification of local protein communities and their organization. Proteins: Struct. Funct. Bioinform. **62**, 800–818 (2006)
3. Hanna, E.M., Zaki, N.: Detecting protein complexes in protein interaction networks using a ranking algorithm with a refined merging procedure. BMC Bioinform. **15**, 204 (2014)
4. Liu, H., Liu, J.: Clustering protein interaction data through chaotic genetic algorithm. In: Simulated Evolution and Learning. Lecture Notes in Computer Science, vol. 4247, pp. 858–864 (2006)
5. Bader, G., Hogue, H.: An automated method for finding molecular com-plexes in large protein-protein interaction networks. BMC Bioinform. **4**, 2 (2003)
6. Adamcsek, B., Palla, G., Farkas, I.J., Derényi, I., Vicsek, T.: CFinder: locating cliques and overlapping modules in biological networks. Bioinformatics **22**, 1021–1023 (2006)
7. Zaki, N., Berengueres, J., Efimov, D.: Detection of protein complexes using a protein ranking algorithm. Proteins: Struct. Funct. Bioinform. **80**(10), 2459–2468 (2012)
8. Guimei, L., Wong, L., Chua, H.N.: Complex discovery from weighted PPI networks. Bioinformatics **25**, 1891–1897 (2009)
9. Nepusz, T., Yu, H., Paccanaro, A.: Detecting overlapping protein complexes in protein-protein interaction networks. Nat. Methods **9**, 471–472 (2012)
10. Zaki, N., Dmitry, D., Berengueres, J.: Protein complex detection using interaction reliability assessment and weighted clustering coefficient. BMC Bioinform. **14**, 1163 (2013)
11. King, A.D., Przulj, N., Jurisica, I.: Protein complex prediction via cost-based clustering. Bioinformatics **17**, 3013–3020 (2004)

12. Enright, A.J., Dongen, S.V., Ouzounis, C.A.: An efficient algorithm for largescale detection of protein families. Nucl. Acids Res. **30**, 1575–1584 (2002)
13. Macropol, K., Can, T., Singh, A.: RRW: repeated random walks on genome-scale protein networks for local cluster discovery. BMC Bioinform. **10**, 283 (2009)
14. Herstein, I.N.: Topics In Algebra. Blaisdell Publishing Company, Waltham (1964)
15. Gavin, A.C., et al.: Functional organization of the yeast proteome by system-atic analysis of protein complexes. Nature **415**, 141–147 (2002)
16. Gavin, A.C., et al.: Proteome survey reveals modularity of the yeast cell machinery. Nature **440**, 631–636 (2006)
17. Leung, H., Xiang, Q., Yiu, S.M., Chin, F.: Predicting protein complexes from PPI data: a core-attachment approach. J. Comput. Biol. **16**, 133–139 (2009)
18. Mewes, H.W., et al.: MIPS: analysis and annotation of proteins from whole genomes. Nucl. Acids Res. **32**, D41–D44 (2004)
19. Pearson, W.R.: Flexible sequence similarity searching with the FASTA3 program package. Methods Mol. Biol. **132**, 185–219 (2000)
20. Altschul, S.F., Madden, T.L., Schäffer, A.A., Zhang, J., Zhang, Z., Miller, W., Lipman, D.J.: Gapped BLAST and PSI-BLAST: a new generation of protein database search programs. Nucl. Acids Res. **25**, 3389–3402 (1997)
21. Aaron J. Mackey, AJ., Haystead, TJ., Pearson, WR.: Algorithms for rapid protein identification with multiple short peptide sequences. Mol. Cell. Proteomics **1**, 139-147 (2002)

Proposal of a New Bioinformatics Pipeline for Metataxonomics in Precision Medicine

Osvaldo Graña-Castro[1,2(✉)], Hugo López-Fernández[2,3,4], Florentino Fdez-Riverola[2,3,4], Fátima Al-Shahrour[1], and Daniel Glez-Peña[2,3,4]

[1] Bioinformatics Unit, Structural Biology Programme, Spanish National Cancer Research Centre (CNIO), C/Melchor Fernández Almagro, 3, 28029 Madrid, Spain
{ograna, falshahrour}@cnio.es

[2] Department of Computer Science, ESEI, University of Vigo, Campus As Lagoas, 32004 Ourense, Spain
{hlfernandez, riverola, dgpena}@uvigo.es

[3] The Biomedical Research Centre (CINBIO), Campus Universitario Lagoas-Marcosende, 36310 Vigo, Spain

[4] SING Research Group, Galicia Sur Health Research Institute (ISS Galicia Sur), SERGAS-UVIGO, Vigo, Spain

Abstract. Microbes are found all over the human body and they have a direct impact on the immune system, metabolism and homeostasis. The homeostatic balance of the intestinal microflora can be broken under certain conditions, a situation known as dysbiosis, which can lead to disease, including certain types of cancer, or even affect a patient response to a therapeutic treatment. Metataxonomics pursues the identification of the bacteria species that are present in biological samples of interest, through the sequencing of the 16S rRNA gene, a highly conserved genetic marker that is present in most prokaryotes. Interactions between the microbiota and the human host are being very relevant in the expansion of precision medicine and cancer research, to better predict the risk of disease and to implement bacteria-directed therapeutics. In order to take metataxonomics to the clinic, efficient bioinformatics pipelines are required, that are flexible and portable, and that are able to classify groups of biological samples according to microbiome diversity. With this objective in mind, we propose a new bioinformatics pipeline to analyze biological samples obtained through NGS of the 16S rRNA gene, doing all the required quality checks and computational calculations. The results obtained with this pipeline are aimed to be interpreted together with host DNA exome or RNA-Seq studies and clinical data, to improve the knowledge about the potential reasons that could lead to disease or to a worst patient treatment response.

Keywords: NGS · 16S rRNA gene · Metataxonomics · Precision medicine

F. Fdez-Riverola et al. (Eds.): PACBB 2019, AISC 1005, pp. 8–15, 2020.
https://doi.org/10.1007/978-3-030-23873-5_2

1 Introduction

The term Microbiota refers to the set of microorganisms that live inside and on humans [1]. Between 500 and 1000 species of bacteria exist in the human body at any one time, corresponding to a ratio of 1.3:1 bacterial cells per human cell [2, 3]. Microbes are found all over the human body, mostly on the external and internal surfaces, including gastrointestinal tract, skin, saliva, oral mucosa, and conjunctiva [4]. The vast majority of the bacteria reside in the colon (large intestine, 10^{11} bacteria/mL content, 400 mL) exceeding all other organs by at least two orders of magnitude, followed by skin ($<10^{11}$ per m2, 1.8m2), Illeum (lower small intestine, 10^{8} bacteria/mL, 400 mL), Duodenum and Jejunum (upper small intestine, 10^{3}–10^{4} bacteria/mL, 400 mL), stomach (10^{3}–10^{4} bacteria/mL, 250–900 mL), saliva (10^{9} bacteria/mL, <100 mL) and dental plaque (10^{11} bacteria/mL, <10 mL).

Genomic studies are focused on the genetic material of a specific organism, while metagenomics is focused on the genetic material of entire communities of organisms [5]. The metagenome of the human microbiome is more variable than the human genome, with only a third of its genes present in a majority of healthy individuals [6]. The microbiome is essential for the maintenance and development of the immune system, metabolism, and homeostasis [7]. Indeed, the homeostatic balance of the intestinal microflora supposes a real benefit to the host, while a drastic imbalance between the beneficial and potentially pathogenic bacteria makes the gut vulnerable to disease. This imbalance is known as dysbiosis [8] and can be of three different types: (i) decrease in the number of beneficial bacteria, (ii) dramatic increase of potentially damaging bacteria and (iii) loss of general bacteria diversity. Dysbiosis has been associated to several diseases including Crohn's disease, ulcerative colitis, obesity, allergic disorders, type I diabetes, autism and cancer. As for cancer, different types have been related to changes in human microbiota, such as gastric carcinoma, colorectal carcinoma, cervical cancer, oral carcinoma and skin cancer [9]. Microbiota can enhance or reduce cancer susceptibility and progression through several ways, such as by modulating inflammation, affecting the genomic stability of host cells and generating metabolites that epigenetically regulate host gene expression [10]. Diet can modulate the composition of the intestinal microbiota, supporting the idea that probiotics and prebiotics could represent effective chemoprevention strategies.

1.1 A Recognized Role of Microbiome in Precision Medicine

The relation between the microbiome and disease could be due to functions coming from a single bacteria species or strain [11], or due to the involvement of multiple bacteria species capable of restoring a wider spectrum of malfunctioning human genes [12]. A recent and very interesting publication by Petrosino [13] points out three articles that illustrate how differential responses to immune checkpoint blockade treatments are related to the patient gut microbiome profile [14–16], two of them showing that a particular set of bacterial species in the gut were associated with successful outcomes in metastatic melanoma patients.

Being able to predict treatment success for patients with a particular disease based on the presence of different bacterial taxa could provide alternative therapeutic approaches.

1.2 Editing the Microbiome is Already a Fact

Petrosino's paper payed careful attention to another publication where the concept of editing the microbiome arised [17]. The authors of this work proposed to treat gastrointestinal inflammatory disease by oral administration of tungstate, which is able to inhibit molybdenum-cofactor-dependent pathways that are essential for Enterobacteriaceae expansion in the inflamed gut [18]. The effect of this inhibition returned microbial diversity to a previous normal state, and the gut inflammation was reduced by up to 90% in experimental mice. The control of the dysbiosis-associated inflammation with this targeted approach shows another potential path towards precision medicine.

Early phase drug trials could benefit from the analysis of the microbiome of responder versus non-responder patients able to influence response to treatment. Furthermore, it could be possible to check if a patient's microbiome would affect the metabolism of certain drugs, providing valuable information for pharmacogenomic profiles [19].

1.3 Metagenomics and Metataxonomics

The term Metagenomics has been splitted into two additional terms: Metagenomics and Metataxonomics [5, 20]. Metagenomics focuses on investigating the genomes of the microbiota through whole-metagenome sequencing (WMS). Metataxonomics relies mostly on the analysis of the 16S rRNA gene sequence, a targeted marker that is amplified to infer a metataxonomic tree.

The gene that encodes 16S rRNA is used as a marker due to its presence in most bacteria, maintaining a practically unchanged function over time, reflected by random changes in its sequence that simply show its evolution [21]. In addition, the 16S rRNA gene has a suitable size for its use as a genetic marker (1500 bp).

2 Pipeline Proposal

We propose a new pipeline in Fig. 1 for the analysis of metataxonomics data, based on the 16S rRNA target gene. This pipeline addresses all the steps that are required to test whether the bacterial population from two groups of biological samples differ in the microbial diversity and its abundance. For instance, between stool samples obtained from groups of responder and non-responder patients to a particular disease treatment. It uses FastQ files [22] with sequenced reads from the different samples and barcode files, as input to Qiime [23]. Qiime validates the sample metadata that contains sample details, such as the sample ID, the sample source, the date of collection, the sample type, and the barcode and primer sequences. Then reads are demultiplexed and quality filtered to reduce noise, removing those that are too short after quality truncation or that contain too many ambiguous base pairs. Quality filtered reads are clustered into

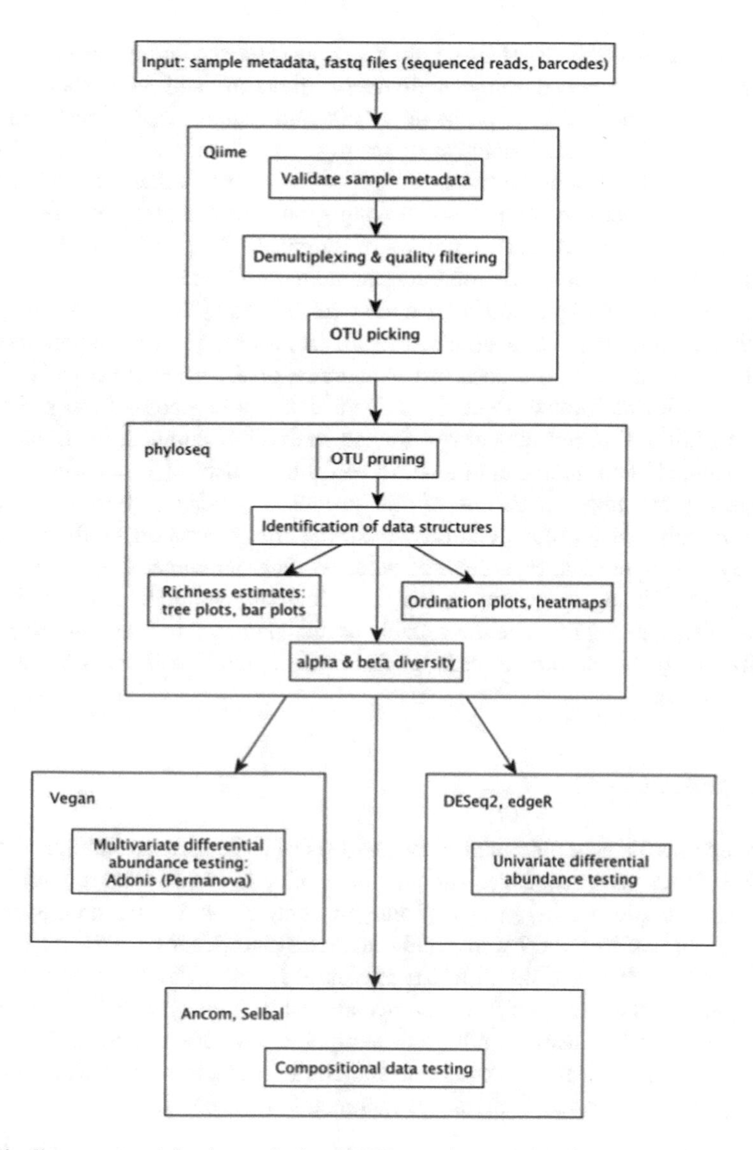

Fig. 1. Pipeline proposed for the analysis of NGS metataxonomics data based on the 16S rRNA gene sequence.

operational taxonomic units (OTUs) based on sequence similarity, usually 97%, to a reference database of known microbial 16S rRNA sequences, such as Greengenes [24] or SILVA [25].

Three Qiime output files are commonly used as input for phyloseq [26]: the taxonomic table, the phylogenetic tree and the OTU table. Phyloseq adds a second level to the pipeline and starts by pruning OTUs that are not present in a required number of samples. Bar plots, tree plots and heatmaps are built to better explore the data. Furthermore, alpha and beta diversity tests are performed. Alpha represents the richness, as a measure of the diversity of OTUs within a sample (observed number of species). Beta corresponds to a 'between samples diversity' measure and is visualized through ordination plots. The Vegan R package [27] is furthermore used to perform a multivariate ANOVA differential abundance testing, based on dissimilarities (PERMANOVA), through the Adonis function. This function tests whether there are significant differences in microbiome composition among groups of samples. Provided that there are significant differences among groups of samples, the pipeline additionally tests which taxa are significantly differentially abundant between sample groups. This univariate differential abundance testing is done with DESeq2 [28] and/or edgeR [29]. It is known that microbiome data is compositional, *i.e.*, a change in the abundance of one taxon induces changes in the observed abundance of the other taxa. Univariate tests (DESeq2, edgeR) performed on compositional data could lead to false positive findings. Two additional algorithms aimed for the analysis of compositional data [30, 31] are also included in our pipeline, in order to verify the reliability of the obtained results.

Regarding the implementation of the proposed pipeline, it must be carefully designed to enhance (i) reproducibility, ensuring the generation of the same results given the same input data, regardless of whenever and wherever they are re-executed, (ii) portability, being easy to install in different computational environments, (iii) scalability, making the pipeline capable of analyzing a larger amount of input data efficiently, and (iv) flexibility, in order to facilitate the modification and/or inclusion of new pipeline steps.

3 Conclusion

We proposed here a new pipeline for the analysis of NGS metataxonomics data, based on the 16S rRNA gene sequence. The pipeline is able to identify the different bacterial species that are present in groups of human samples, and their abundance in the samples. It can perform multivariate and univariate significant tests, to better show the difference in abundance of the different microbial species. Finally, it applies two different algorithms that are specific for compositional data analysis in order to verify the results and avoid false positive findings. The pipeline results are aimed to be interpreted together with human host/patient genome and clinical data, to try to find out early signs of disease and assist in the decision for therapeutic treatments.

Acknowledgments. The SING group thanks the CITI (Centro de Investigación, Transferencia e Innovación) from the University of Vigo for hosting its IT infrastructure. This work was partially supported by the Consellería de Educación, Universidades e Formación Profesional (Xunta de Galicia) under the scope of the strategic funding ED431C2018/55-GRC Competitive Reference Group, and by the Plataforma de Bioinformática from the Instituto de Salud Carlos III (PT17/0009/0011).

References

1. Turnbaugh, P.J., Ley, R.E., Hamady, M., Fraser-Liggett, C.M., Knight, R., Gordon, J.I.: The human microbiome project. Nature **449**, 804–810 (2007)
2. Sender, R., Fuchs, S., Milo, R.: Are we really vastly outnumbered? Revisiting the ratio of bacterial to host cells in humans. Cell **164**, 337–340 (2016)
3. Gilbert, J.A., Blaser, M.J., Caporaso, J.G., Jansson, J.K., Lynch, S.V., Knight, R.: Current understanding of the human microbiome. Nat. Med. **24**, 392–400 (2018)
4. Sender, R., Fuchs, S., Milo, R.: Revised estimates for the number of human and bacteria cells in the body. PLoS Biol. **14**, e1002533 (2016)
5. Aguiar-Pulido, V., Huang, W., Suarez-Ulloa, V., Cickovski, T., Mathee, K., Narasimhan, G.: Metagenomics, metatranscriptomics, and metabolomics approaches for microbiome analysis. Evol. Bioinform. Online **12**, 5–16 (2016)
6. Lloyd-Price, J., Abu-Ali, G., Huttenhower, C.: The healthy human microbiome. Genome Med. **8**, 51 (2016)
7. Kim, B.-S., Jeon, Y.-S., Chun, J.: Current status and future promise of the human microbiome. Pediatr. Gastroenterol. Hepatol. Nutr. **16**, 71–79 (2013)
8. DeGruttola, A.K., Low, D., Mizoguchi, A., Mizoguchi, E.: Current understanding of dysbiosis in disease in human and animal models. Inflamm. Bowel Dis. **22**, 1137–1150 (2016)
9. Banerjee, J., Mishra, N., Dhas, Y.: Metagenomics: a new horizon in cancer research. Meta Gene **5**, 84–89 (2015)
10. Bultman, S.J.: Emerging roles of the microbiome in cancer. Carcinogenesis **35**, 249–255 (2014)
11. Buffington, S.A., Di Prisco, G.V., Auchtung, T.A., Ajami, N.J., Petrosino, J.F., Costa-Mattioli, M.: Microbial reconstitution reverses maternal diet-induced social and synaptic deficits in offspring. Cell **165**, 1762–1775 (2016)
12. Gevers, D., Kugathasan, S., Denson, L.A., Vazquez-Baeza, Y., Van Treuren, W., Ren, B., Schwager, E., Knights, D., Song, S.J., Yassour, M., Morgan, X.C., Kostic, A.D., Luo, C., Gonzalez, A., McDonald, D., Haberman, Y., Walters, T., Baker, S., Rosh, J., Stephens, M., Heyman, M., Markowitz, J., Baldassano, R., Griffiths, A., Sylvester, F., Mack, D., Kim, S., Crandall, W., Hyams, J., Huttenhower, C., Knight, R., Xavier, R.J.: The treatment-naive microbiome in new-onset Crohn's disease. Cell Host Microbe **15**, 382–392 (2014)
13. Petrosino, J.F.: The microbiome in precision medicine: the way forward. Genome Med. **10**, 12 (2018)
14. Gopalakrishnan, V., Spencer, C.N., Nezi, L., Reuben, A., Andrews, M.C., Karpinets, T.V., Prieto, P.A., Vicente, D., Hoffman, K., Wei, S.C., Cogdill, A.P., Zhao, L., Hudgens, C.W., Hutchinson, D.S., Manzo, T., Petaccia de Macedo, M., Cotechini, T., Kumar, T., Chen, W. S., Reddy, S.M., Szczepaniak Sloane, R., Galloway-Pena, J., Jiang, H., Chen, P.L., Shpall, E.J., Rezvani, K., Alousi, A.M., Chemaly, R.F., Shelburne, S., Vence, L.M., Okhuysen, P. C., Jensen, V.B., Swennes, A.G., McAllister, F., Marcelo Riquelme Sanchez, E., Zhang, Y., Le Chatelier, E., Zitvogel, L., Pons, N., Austin-Breneman, J.L., Haydu, L.E., Burton, E.M., Gardner, J.M., Sirmans, E., Hu, J., Lazar, A.J., Tsujikawa, T., Diab, A., Tawbi, H., Glitza, I. C., Hwu, W.J., Patel, S.P., Woodman, S.E., Amaria, R.N., Davies, M.A., Gershenwald, J.E., Hwu, P., Lee, J.E., Zhang, J., Coussens, L.M., Cooper, Z.A., Futreal, P.A., Daniel, C.R., Ajami, N.J., Petrosino, J.F., Tetzlaff, M.T., Sharma, P., Allison, J.P., Jenq, R.R., Wargo, J. A.: Gut microbiome modulates response to anti-PD-1 immunotherapy in melanoma patients. Science **359**, 97–103 (2018)

15. Matson, V., Fessler, J., Bao, R., Chongsuwat, T., Zha, Y., Alegre, M.-L., Luke, J.J., Gajewski, T.F.: The commensal microbiome is associated with anti-PD-1 efficacy in metastatic melanoma patients. Science **359**, 104–108 (2018)
16. Routy, B., Le Chatelier, E., Derosa, L., Duong, C.P.M., Alou, M.T., Daillere, R., Fluckiger, A., Messaoudene, M., Rauber, C., Roberti, M.P., Fidelle, M., Flament, C., Poirier-Colame, V., Opolon, P., Klein, C., Iribarren, K., Mondragon, L., Jacquelot, N., Qu, B., Ferrere, G., Clemenson, C., Mezquita, L., Masip, J.R., Naltet, C., Brosseau, S., Kaderbhai, C., Richard, C., Rizvi, H., Levenez, F., Galleron, N., Quinquis, B., Pons, N., Ryffel, B., Minard-Colin, V., Gonin, P., Soria, J.-C., Deutsch, E., Loriot, Y., Ghiringhelli, F., Zalcman, G., Goldwasser, F., Escudier, B., Hellmann, M.D., Eggermont, A., Raoult, D., Albiges, L., Kroemer, G., Zitvogel, L.: Gut microbiome influences efficacy of PD-1-based immunotherapy against epithelial tumors. Science **359**, 91–97 (2018)
17. Zhu, W., Winter, M.G., Byndloss, M.X., Spiga, L., Duerkop, B.A., Hughes, E.R., Buttner, L., de Lima Romao, E., Behrendt, C.L., Lopez, C.A., Sifuentes-Dominguez, L., Huff-Hardy, K., Wilson, R.P., Gillis, C.C., Tukel, C., Koh, A.Y., Burstein, E., Hooper, L.V., Baumler, A. J., Winter, S.E.: Precision editing of the gut microbiota ameliorates colitis. Nature **553**, 208–211 (2018)
18. Winter, S.E., Winter, M.G., Xavier, M.N., Thiennimitr, P., Poon, V., Keestra, A.M., Laughlin, R.C., Gomez, G., Wu, J., Lawhon, S.D., Popova, I.E., Parikh, S.J., Adams, L.G., Tsolis, R.M., Stewart, V.J., Baumler, A.J.: Host-derived nitrate boosts growth of E. coli in the inflamed gut. Science **339**, 708–711 (2013)
19. Spanogiannopoulos, P., Bess, E.N., Carmody, R.N., Turnbaugh, P.J.: The microbial pharmacists within us: a metagenomic view of xenobiotic metabolism. Nat. Rev. Microbiol. **14**, 273–287 (2016)
20. Marchesi, J.R., Ravel, J.: The vocabulary of microbiome research: a proposal. Microbiome **3**, 31 (2015)
21. Janda, J.M., Abbott, S.L.: 16S rRNA gene sequencing for bacterial identification in the diagnostic laboratory: pluses, perils, and pitfalls. J. Clin. Microbiol. **45**, 2761–2764 (2007)
22. Cock, P.J.A., Fields, C.J., Goto, N., Heuer, M.L., Rice, P.M.: The sanger FASTQ file format for sequences with quality scores, and the Solexa/Illumina FASTQ variants. Nucleic Acids Res. **38**, 1767–1771 (2010)
23. Caporaso, J.G., Kuczynski, J., Stombaugh, J., Bittinger, K., Bushman, F.D., Costello, E.K., Fierer, N., Pena, A.G., Goodrich, J.K., Gordon, J.I., Huttley, G.A., Kelley, S.T., Knights, D., Koenig, J.E., Ley, R.E., Lozupone, C.A., McDonald, D., Muegge, B.D., Pirrung, M., Reeder, J., Sevinsky, J.R., Turnbaugh, P.J., Walters, W.A., Widmann, J., Yatsunenko, T., Zaneveld, J., Knight, R.: QIIME allows analysis of high-throughput community sequencing data. Nat. Methods **7**, 335–336 (2010)
24. DeSantis, T.Z., Hugenholtz, P., Larsen, N., Rojas, M., Brodie, E.L., Keller, K., Huber, T., Dalevi, D., Hu, P., Andersen, G.L.: Greengenes, a chimera-checked 16S rRNA gene database and workbench compatible with ARB. Appl. Environ. Microbiol. **72**, 5069–5072 (2006)
25. Yilmaz, P., Parfrey, L.W., Yarza, P., Gerken, J., Pruesse, E., Quast, C., Schweer, T., Peplies, J., Ludwig, W., Glockner, F.O.: The SILVA and "All-species Living Tree Project (LTP)" taxonomic frameworks. Nucleic Acids Res. **42**, D643–D648 (2014)
26. McMurdie, P.J., Holmes, S.: phyloseq: an R package for reproducible interactive analysis and graphics of microbiome census data. PLoS ONE **8**, e61217 (2013)
27. Oksanen, J., Blanchet, G.F., Kindt, R., Legendre, P., Minchin, P.R., O'Hara, R.B., Simpson, G.L., Solymos, P., Stevens, M.H., Wagner, H.: vegan: Community Ecology Package, R package version 2.3-0 (2015)

28. Love, M.I., Huber, W., Anders, S.: Moderated estimation of fold change and dispersion for RNA-seq data with DESeq2. Genome Biol. **15**, 550 (2014)
29. Robinson, M.D., McCarthy, D.J., Smyth, G.K.: edgeR: a Bioconductor package for differential expression analysis of digital gene expression data. Bioinformatics **26**, 139–140 (2010)
30. Mandal, S., Van Treuren, W., White, R.A., Eggesbo, M., Knight, R., Peddada, S.D.: Analysis of composition of microbiomes: a novel method for studying microbial composition. Microb. Ecol. Health Dis. **26**, 27663 (2015)
31. Rivera-Pinto, J., Egozcue, J.J., Pawlowsky-Glahn, V., Paredes, R., Noguera-Julian, M., Calle, M.L.: Balances: a new perspective for microbiome analysis. mSystems **3**(4), e00053-18 (2018)

Systems Toxicology Approach to Unravel Early Indicators of Squamous Cell Carcinoma Rate in Rat Nasal Epithelium Induced by Formaldehyde Exposure

Florian Martin(✉), Marja Talikka, Julia Hoeng, and Manuel C. Peitsch

PMI R&D, Philip Morris Products S.A., Neuchâtel, Switzerland
{florian.martin,marja.talikka,julia.hoeng, manuel.peitsch}@pmi.com

Abstract. Causal biological network models consisting of multiple biological pathways involved in a given biological process can serve to contextualize gene expression changes and unravel key mechanisms responsible for those changes. The transcriptomic data from the respiratory nasal epithelium (RNE) of rats exposed to formaldehyde have been investigated using such causal biological network models. The resulting association between the biological impact assessed by network perturbation and the squamous cell carcinoma rate in the RNE after two years has been further investigated to gain mechanistic insights. A detailed node-level investigation revealed that while similar network models were impacted across exposure doses, the directionality of the effect was opposite for the lowest doses compared to high doses. In particular, NF-κB was inferred to be upregulated in response to the two higher doses and downregulated in response to the lower doses in the context of the epithelial innate immune activation network model. This highlighted a dose threshold indicative of a long-term biphasic effect of formaldehyde exposure leading to carcinogenicity. The presented approach could be used to establish the mechanism of action or grouping of compounds based on impacted regions in the network models.

Keywords: Gene expression · Network biology · Toxicology

1 Introduction

The systems approach is gaining popularity in toxicology in the awareness that apical endpoints only give a partial view of the effects of exposure to substances and have little predictive value [1, 2]. Therefore, the inclusion of large scale datasets, which is an essential part of systems toxicology, is imperative to gain a more global understanding of the biological responses caused by substance exposure.

The analysis of these large scale datasets requires sophisticated bioinformatics and computational modelling approaches. A frequently used analysis method is to overlay the molecular changes onto biological pathway maps [3], which provides insights into

F. Fdez-Riverola et al. (Eds.): PACBB 2019, AISC 1005, pp. 16–24, 2020.
https://doi.org/10.1007/978-3-030-23873-5_3

how biological pathways are affected by substance exposure. Gene expression data can also be used to infer gene regulatory networks to answer toxicological questions and unravel exposure effects [4].

Causal biological network models consisting of multiple biological pathways involved in a given biological process can be built from literature using computable languages [5]. Beyond the forward assumption (used in pathway analysis and gene regulatory networks), by which changes in gene expression imply that the activity of the gene product is altered accordingly, the algorithm designed to be used in conjunction with causal biological network models uses the gene expression data to infer the activity of a node in the network [6]. In addition to the mechanistic information, network scoring provides a quantitative measure of the exposure impact. To date, several crowd-verified models for pulmonary, vascular, and brain biology have been built and used in large-scale data interpretation [7].

To showcase the causal biological network-based approach, we have previously reanalyzed a public transcriptomic dataset from the respiratory nasal epithelium (RNE) of rats exposed to increasing doses of formaldehyde (FA) [8, 9]. While FA is made in the human body, chronic exposure to FA concentrations used in industry causes adverse effects [10] and triggers tumorigenesis in the RNE of rats after prolonged exposure [11, 12]. Our earlier work showed a good correlation of the biological impact assessed by network perturbation perturbations at 13 weeks of exposure and the carcinoma rate at two years [8]. To expand on this analysis, we have focused on the mechanistic explanation of the carcinoma rate in response to different FA doses after 13 weeks of exposure.

2 Methods

2.1 Causal Biological Network Models

The building and uses of the causal biological network models have been described in various publications and use cases [7]. Briefly, the network models were built using the Biological Expression Language (BEL), which converts scientific results (i.e., molecular cause-and-effect relationships) into statements that can be computed into interconnected network models for visualization of the biology as well as scoring with high-throughput data [6, 13, 17]. The biological entities in the network models are called "nodes" and are connected by edges (i.e., the causal relationships between the molecular entities). The network model suite built to describe the biological processes of the respiratory system consists of five network families (cell fate [CFA], cell proliferation [CPR], cell stress [CST], inflammatory processes [IPN], and tissue repair and angiogenesis [TRA], each containing several individual process-specific network models (e.g., cell cycle within CPR). All of these causal biological network models are available for browsing and download at causalbionet.com [13, 17].

2.2 Network Perturbation Amplitude and Biological Impact Factor

The transcriptomic dataset (GSE23179) used in this analysis was derived from an FA inhalation study conducted in rats and described in [9]. Raw data files were processed with the custom Chip Description File environment Rat2302_Rn_ENTREZG v19.0 [20] and normalized using frozen robust microarray analysis (fRMA) [21]. The normalization vector needed for fRMA was created using a set of 1,023 microarray rat samples from 11 tissues and the R package frmaTools version 1.18.0. Quality controls, including log-intensities, normalized unscaled standard error, relative log expression (RLE), and median absolute value RLE, as well as pseudo-images and raw image plots, were performed with the affyPLM package (Bioconductor, USA) [22].

The network perturbation amplitude (NPA) method was previously reported [6, 8, 14]. Briefly, the methodology aims at contextualizing transcriptome profiles (exposed vs. non-exposed) by combining the alteration of gene expression into differential node values (i.e., one value for each node of the causal network model). Literature-derived and experimental information supporting the relationship between network nodes and the expression of certain genes is included. Thus, a transcriptome profile can be used to computationally predict the activity of network nodes by a fitting procedure inferring the values that best satisfy the directionality of the causal relationships contained in the network model (e.g., positive or negative signs). The node values are in turn summarized in to a single perturbation score, the NPA. In addition to a confidence interval accounting for the experimental variation (and the associated *p*-values), companion statistics, derived to inform the specificity of the NPA score to the biology described in the network models, are reported as *O and K* if their *p*-values fall below the threshold of significance (0.05). A network is therefore considered to be significantly impacted by exposure if the three values (the *p*-value for experimental variation, *O, and K* statistics) are below 0.05. In addition to the impact/perturbation scores at the levels of network, the effects of exposure were further quantified as a system-wide metric for biological impact, the biological impact factor (BIF) [8, 14, 23]. This positive value of BIF summarizes the impacts of the exposure on the cellular system into a single number, thus enabling a simple and high-level evaluation of the treatment effects across various doses.

2.3 Leading Node Analysis

While the NPA score is positive by definition, the differential backbone values are informative of the direction of changes in the network. The leading nodes of a significant network are defined to be the nodes that contribute the most, up to 80%, to the NPA value [6]. When interpreting the network response to the exposure, and in order to avoid investigating the numerous nodes of a network, the leading node analysis focuses on the most impacted areas of the network.

3 Results

3.1 Network Response in the Rat RNE in Response to FA Exposure for 13 Weeks

We previously analyzed the transcriptomic data from the RNE of rats exposed to FA [9] using the causal biological network models in [8]. We then focused on the assessment of the network response to FA exposure across the time points [8], and in this work, we sought to gain mechanistic insight into the dose effect regarding carcinoma rate. Moreover, the network models have since undergone crowd verification and have been updated [7] with the new versions (1.3) available in the Causal Biological Network Database (causalbionet.com) [13, 17]. Finally, a new network model family describing the biological processes in tissue repair and angiogenesis was included in the respiratory model suite [24]. The transcriptomic data from the FA-exposed rat RNE

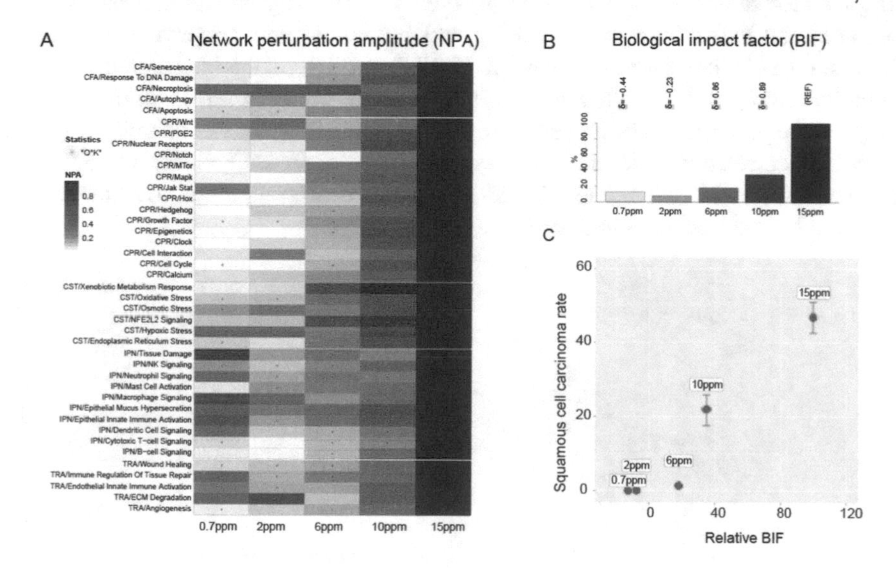

Fig. 1. NPA and BIF in the RNE of rats exposed to FA. (A) The data at one day post-exposure was scored against causal biological network models, as described in [6]. A network is considered as perturbed if, in addition to the significance of the NPA score with respect to the experimental variation, the two companion statistics (O and K), derived to inform on the specificity of the NPA score with respect to the biology described in the network, are significant. Symbol legend: *: O and K statistic *p*-values below 0.05 and NPA significant with respect to the experimental variation. (B) BIF values were derived for each contrast (e.g., treatment vs. control) from the aggregated NPA scores for networks applicable to the set of contrasts. The treatment comparison showing the highest perturbation is set to 100%. For each treatment comparison, the δ value on the top of the bar (–1 to 1) indicates how similar the underlying network perturbations (with respect to impacted nodes and directionality) are with respect to the reference (i.e., REF). A δ value of 1 indicates that all the networks are perturbed by the same mechanisms. (C) The correlation of BIF at 13 weeks and carcinoma rate at two years. Carcinoma rate data was obtained from [12]. Ppm, parts per million; CFA, cell fate; CPR, cell proliferation; CST, cell stress; IPN, inflammatory processes; TRA, tissue repair and angiogenesis.

at the 13-week time point were scored against the updated suite of causal biological network models. While many network models were impacted by all FA concentrations, the strongest impact was observed in response to the 15 ppm dose (Fig. 1A). The response to DNA damage and the epigenetics network models were significantly impacted only by 6, 10, and 15 ppm FA.

This was further demonstrated in the aggregated NPA score of all network models, the BIF (Fig. 1B). The results were well aligned with those published using the previous network versions [8]. The correlation between the BIF score and the squamous cell carcinoma rate in the RNE after two years [12] (Fig. 1C) shows that with 0.7 and 2 ppm FA concentrations, the BIF remained low at 13 weeks, with virtually no carcinomas at two years [9, 12].

3.2 Mechanistic Interpretation of the FA Effect at 13 Weeks

To gain mechanistic insight into the dose effect regarding carcinoma rate, we next compared the network impact at family and individual network levels between the different doses at the 13-week time point. For 0.7 and 2 ppm, the impacted mechanisms primarily involved IPN, whereas the higher doses (leading to carcinoma) mainly triggered the perturbation of processes contained in the CFA network family (Fig. 2A).

A high correlation at the network level could be observed between the 6, 10, and 15 ppm doses and between the 0.7 and 2 ppm doses. The high and low doses correlated poorly at the individual network level (Fig. 2B).

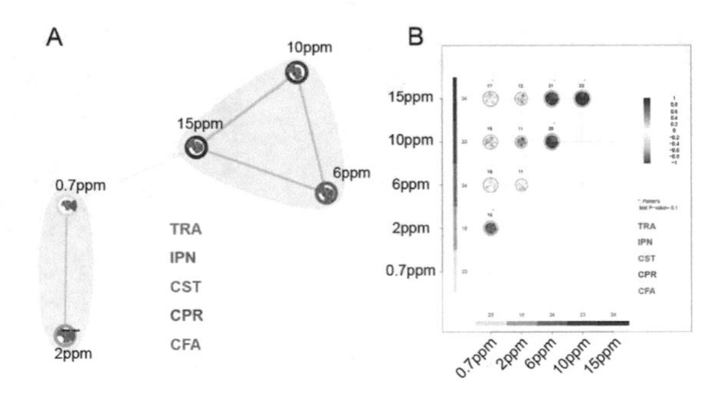

Fig. 2. Comparison of biological mechanisms between the different FA concentrations. (A) Comparison at the network family level. Comparisons of the contrasts based on the Manhattan distance between the leading node contributions across all the significant subnetworks. Clustering is performed in two stages using Affine Propagation [25]. The algorithm was applied once and re-applied between the exemplars (representative point of each cluster). Edges relate nodes within the same cluster. (B) Correlations of differential network values for each network. CFA, cell fate; CPR, cell proliferation; CST, cell stress; IPN, inflammatory processes; TRA, tissue repair and angiogenesis.

The poor correlation between the biology that was impacted in the rat RNE in response to the low and high doses of FA prompted us to look into the perturbation of the individual network models in more detail. While the NPA scoring gives insight into the impact of exposures on the network families and the individual network models within, the leading node analysis (described in the Methods section) is needed to locate the exact regions of the network models that are impacted as well as the directionality of the impact. We extracted the most important perturbed nodes in the network models (leading nodes), with the graphs illustrating the NPA value for each individual node in response to the five FA doses compared with the controls. The closer look at the epithelial innate immune activation network model showed the opposite regulation of nuclear factor kappa B (NF-κB) in response to the low and high doses of FA. The mechanistic investigation of the immune regulation of the tissue repair network model further demonstrated that while many inflammatory mediators, such as Tnf, Ifng, Stats, Mmp12, and chemokines, as well as the activation of macrophages and neutrophils, were inferred to be increased in response to the high FA doses, these mechanisms were inferred to be downregulated in response to the lower FA doses (Fig. 3). These examples and many others (not shown) underscore the importance of the leading node analysis; the impact observed at network NPA and BIF levels can be either positive or negative. Moreover, the analyses highlighted the underlying inferred molecular changes responsible for the network impact.

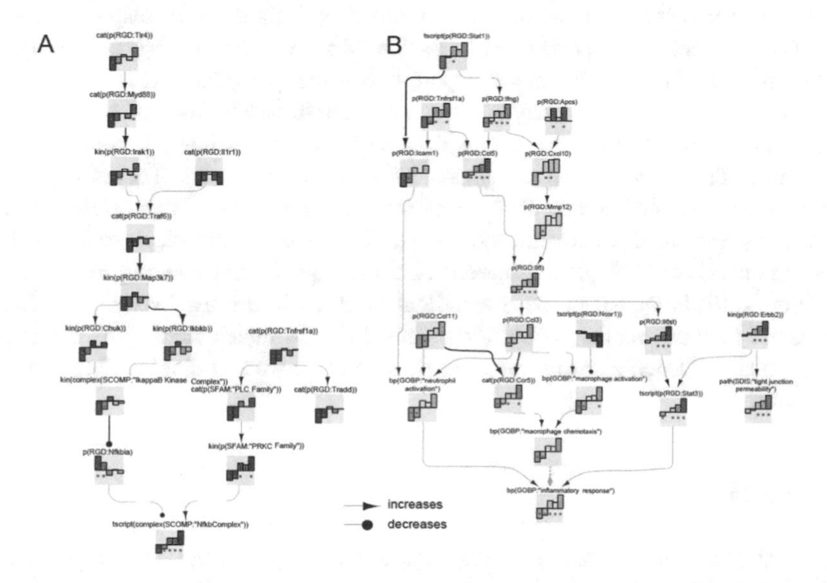

Fig. 3. Extracts of the epithelial innate immune activation (A) and the immune regulation of tissue repair (B) network models with connected leading nodes common for the 13-week time point. The backbone NPA values with directionalities of inferred regulation are shown as bar graphs for each node. Orange/red bars indicate inferred upregulation, and blue bars indicate inferred downregulation. The green asterisk indicates that the node is a leading node. The entire networks are available for view and download at causalbionet.com, and the vocabulary for the BEL is provided at http://www.openbel.org/.

4 Discussion

The scoring of causal biological network models with transcriptomic data provides a sensitive and quantitative means to assess exposure effects. Here, we have extended our previous work on the rat FA dataset and focused on the 13-week time point to find mechanistic differences between the low and high doses that may explain the threshold for carcinoma development in the RNE.

The overall correlation between the low and high doses was poor at both the network family and individual network levels, and the first indication of the more tumorigenic potential of the three higher doses became apparent from the impact of the response on DNA repair and epigenetics network models. The more detailed node-level investigation revealed that while similar network models were impacted across the doses, the directionality of the effect was, in some cases, opposite. In particular, NF-κB was inferred to be upregulated in response to the two higher doses and downregulated in response to the lower doses in the context of the epithelial innate immune activation network model. NF-κB is involved in many aspects of cancer biology and inflammation and could certainly be the defining mechanistic threshold for carcinoma at a later time point [26]. The immune regulation of tissue repair network model replicated the indication that the high doses of FA induced an inflammatory response in the rat RNE after 13 weeks of exposure, whereas the two lower doses triggered an apparent downregulation of these pathways compared with the unexposed rats.

In conclusion, we could capture the shift in impacted biological mechanisms between the 2 and 6 ppm FA doses in a quantitative manner, demonstrating the opportunities the network approach presents for cancer bioassays. The approach could also be used to establish the mechanism of action or grouping of compounds based on impacted regions in the network models. In fact, the possibilities are far-reaching, extending outside toxicological research. The same questions regarding mode of action and adverse effects appear in the biomedical field, both in drug discovery and safety. Finally, the network models could aid in mechanistic understanding of disease processes, which is imperative in biomarker discovery as well as disease diagnostics and prognostics.

References

1. Council NR. Toxicity testing in the 21st century: a vision and a strategy. National Academies Press, 2007
2. Sturla, S.J., Boobis, A.R., FitzGerald, R.E., et al.: Systems toxicology: from basic research to risk assessment. Chem. Res. Toxicol. **27**, 314–329 (2014)
3. Alexander-Dann, B., Pruteanu, L.L., Oerton, E., et al.: Developments in toxicogenomics: understanding and predicting compound-induced toxicity from gene expression data. Mol. Omics **14**, 218–236 (2018)
4. Emmert-Streib, F., Dehmer, M., Haibe-Kains, B.: Gene regulatory networks and their applications: understanding biological and medical problems in terms of networks. Front. Cell Dev. Biol. **2**, 38 (2014)

5. Szostak, J., Ansari, S., Madan, S., et al.: Construction of biological networks from unstructured information based on a semi-automated curation workflow. Database **2015**, bav057 (2015)
6. Martin, F., Sewer, A., Talikka, M., Xiang, Y., Hoeng, J., Peitsch, M.C.: Quantification of biological network perturbations for mechanistic insight and diagnostics using two-layer causal models. BMC Bioinform. **15**, 238 (2014)
7. Talikka, M., Bukharov, N., Hayes, W.S., et al.: Novel approaches to develop community-built biological network models for potential drug discovery. Expert Opin. Drug Discov. **12**, 849–857 (2017)
8. Hoeng, J., Talikka, M., Martin, F., et al.: Toxicopanomics: applications of genomics, transcriptomics, proteomics, and lipidomics in predictive mechanistic toxicology. In: Hayes' Principles and Methods of Toxicology, p. 322–359. CRC Press (2014)
9. Andersen, M.E., Clewell 3rd, H.J., Bermudez, E., et al.: Formaldehyde: integrating dosimetry, cytotoxicity, and genomics to understand dose-dependent transitions for an endogenous compound. Toxicol. Sci. **118**, 716–731 (2010)
10. Bernstein, R.S., Stayner, L.T., Elliott, L.J., Kimbrough, R., Falk, H., Blade, L.: Inhalation exposure to formaldehyde: an overview of its toxicology, epidemiology, monitoring, and control. Am. Ind. Hyg. Assoc. J. **45**, 778–785 (1984)
11. Kerns, W.D., Pavkov, K.L., Donofrio, D.J., Gralla, E.J., Swenberg, J.A.: Carcinogenicity of formaldehyde in rats and mice after long-term inhalation exposure. Cancer Res. **43**, 4382–4392 (1983)
12. Monticello, T.M., Swenberg, J.A., Gross, E.A., et al.: Correlation of regional and nonlinear formaldehyde-induced nasal cancer with proliferating populations of cells. Cancer Res. **56**, 1012–1022 (1996)
13. Boué, S., Talikka, M., Westra, J.W., et al.: Causal biological network database: a comprehensive platform of causal biological network models focused on the pulmonary and vascular systems. Database **2015**, bav030 (2015)
14. Hoeng, J., Talikka, M., Martin, F., et al.: Case study: the role of mechanistic network models in systems toxicology. Drug Discov. Today **19**, 183–192 (2014)
15. Kogel, U., Titz, B., Schlage, W.K., et al.: Evaluation of the tobacco heating system 2.2. Part 7: systems toxicological assessment of a mentholated version revealed reduced cellular and molecular exposure effects compared with mentholated and non-mentholated cigarette smoke. Regul. Toxicol. Pharmacol. **81**, S123–S138 (2016)
16. Phillips, B., Veljkovic, E., Boué, S., et al.: An 8-month systems toxicology inhalation/cessation study in Apoe−/− mice to investigate cardiovascular and respiratory exposure effects of a candidate modified risk tobacco product, THS 2.2, compared with conventional cigarettes. Toxicol. Sci. **149**, 411–432 (2015)
17. Talikka, M., Boue, S., Schlage, W.K.: Causal biological network database: a comprehensive platform of causal biological network models focused on the pulmonary and vascular systems. In: Hoeng, J., Peitsch, M.C. (eds.) Computational Systems Toxicology. MPT, pp. 65–93. Springer, New York (2015). https://doi.org/10.1007/978-1-4939-2778-4_3
18. Wong, E.T., Kogel, U., Veljkovic, E., et al.: Evaluation of the Tobacco Heating System 2.2. Part 4: 90-day OECD 413 rat inhalation study with systems toxicology endpoints demonstrates reduced exposure effects compared with cigarette smoke. Regul. Toxicol. Pharmacol. **81**, S59–S81 (2016)
19. Zanetti, F., Sewer, A., Scotti, E., et al.: Assessment of the impact of aerosol from a potential modified risk tobacco product compared with cigarette smoke on human organotypic oral epithelial cultures under different exposure regimens. Food Chem. Toxicol. **115**, 148–169 (2018)

20. Dai, M., Wang, P., Boyd, A.D., et al.: Evolving gene/transcript definitions significantly alter the interpretation of GeneChip data. Nucl. Acids Res. **33**, e175 (2005)
21. McCall, M.N., Bolstad, B.M., Irizarry, R.A.: Frozen robust multiarray analysis (fRMA). Biostatistics **11**, 242–253 (2010)
22. Bolstad, B.M., Irizarry, R.A., Åstrand, M., Speed, T.P.: A comparison of normalization methods for high density oligonucleotide array data based on variance and bias. Bioinformatics **19**, 185–193 (2003)
23. Thomson, T.M., Sewer, A., Martin, F., et al.: Quantitative assessment of biological impact using transcriptomic data and mechanistic network models. Toxicol. Appl. Pharmacol. **272**, 863–878 (2013)
24. Park, J., Schlage, W., Frushour, B., Talikka, M., Toedter, G.: Construction of a computable network model of tissue repair and angiogenesis in the lung. J. Clin. Toxicol. **S12**, 002 (2013). https://doi.org/10.4172/2161-0495.S12-002
25. Frey, B.J., Dueck, D.: Clustering by passing messages between data points. Science **315**, 972–976 (2007)
26. Karin, M.: NF-κB as a critical link between inflammation and cancer. Cold Spring Harb. Persp. Biol. **1**, a000141 (2009)

Moment Vector Encoding of Protein Sequences for Supervised Classification

Haneen Altartouri(✉) and Tobias Glasmachers

Institute for Neural Computation, Ruhr-University Bochum, Bochum, Germany
{haneen.altartouri,tobias.glasmachers}@ini.rub.de

Abstract. Automated prediction of biological attributes of protein sequences with machine learning methods depends on a well-suited protein representation. A central challenge is to represent variable-length sequences as fixed-length feature vectors. In this paper we introduce a new approach for representing the protein sequences as a fixed length vector based on statistical moments applied directly to the values of physicochemical properties of amino acids. The results show that this approach of encoding gives higher prediction accuracy on four benchmarks compared to the previous approaches that applied moments of complex descriptors extracted from the physicochemical properties, and even better than the PseAAC encoding method. The best results are achieved by removing highly correlated features with principal component analysis.

Keywords: Moment vector · Protein sequences · Physicochemical properties

1 Introduction

With the increasing number of biological sequences, prediction of biological attributes such as function, structure and localization, based on protein sequences is gaining more attention [18]. Several researchers have focused on using different computational approaches to solve various protein prediction problems in a faster and more cost-effective manner, ideally avoiding lab experiments altogether.

Most machine learning models are restricted to inputs of an a-priori fixed dimensions. One of the most critical challenges in predicting a protein's attributes is how to effectively represent a variable-length sequence for processing with such a predictive model. Some model types like recurrent networks can handle variable-length sequences directly, however, in this study we focus on support vector machines since they seem to perform best on this type of problem [12].

Recently, several researchers have explored the utility of using moment vectors to represent protein sequences. The main motivation is that moment vectors can represent variable length sequences with a fixed-length descriptor.

F. Fdez-Riverola et al. (Eds.): PACBB 2019, AISC 1005, pp. 25–35, 2020.
https://doi.org/10.1007/978-3-030-23873-5_4

Yau et al. [21] proved that the correspondence between a protein sequence and its moment vector is one-to-one, where each moment represents information from the whole sequence. Moments can be defined for any sequence of numbers. To represent protein sequences as moment vectors, some approaches depend on the character representation of amino acids [22], while others use descriptors derived from the native values of Physicochemical properties (PCPs) [20, 21].

Yau et al. [21] used moment vectors to improve the clustering of one kind of proteins (kinase C family). They used scaled descriptors of the hydrophobicity values of amino acids to represent the sequences. These scaled values were used to compute the moment vectors of the sequences. They illustrated that this method can specify the similarity of two protein sequences.

On the other hand, Sun et al. [20] used six PCPs of amino acids and the relationship between them in order to create an adjacency matrix for each amino acid. Based on the eigenvalues of these adjacency matrices they got a graph energy value for each amino acid, which is thought of as an aggregated descriptor. The energy values were used to generate moment vectors to represent protein sequences. They assessed the similarities and dissimilarities of several typical datasets. The results showed that this method is better suited than previous approaches for studying similarities between sequences.

Most of the existing works on representing protein sequences with moment vectors focuses on solving clustering problems. Moments are not computed for the native values of PCPs, but from derived quantities. The results are rarely compared across a wide range of sequence prediction problems.

In this paper we go beyond the state of the art in several ways. First of all, we explore the power of moment-based representations for supervised classification. Instead of going through involved constructions, we apply moments directly to PCP values, for different promising sets of PCP descriptors. We consider the whole processing pipeline of supervised learning, namely data representation, feature selection and extraction, and the final classification step using supervised machine learning methods. The evaluation is based on four diverse protein sequence prediction problems.

The remainder of this paper is organized as follows: In the next section we describe the proposed encoding method using moment vectors. Then we briefly introduce the SVM algorithm. We assess our techniques experimentally and evaluate the results. We close with our conclusions.

2 Moment Vectors of PCPs

A protein sequence can be represented as a curve or graph of the consecutive PCP values of its amino acids. Figure 1a illustrates the graphical representation of the HIV-1 sequence AEAMSQVT using hydrophobicity values (Tanford, 1980) [6] of amino acids.

In Fig. 1a, the curve is formed by the sequence of points $(1, y_1)$, $(2, y_2), (3, y_3), \ldots, (n, y_n)$, where the horizontal axis represents the positions of the amino acid within the sequence and the vertical axis represents the corresponding PCP value. The curve points $(1, y_1), (2, y_2), (3, y_3), \ldots, (n, y_n)$ can be

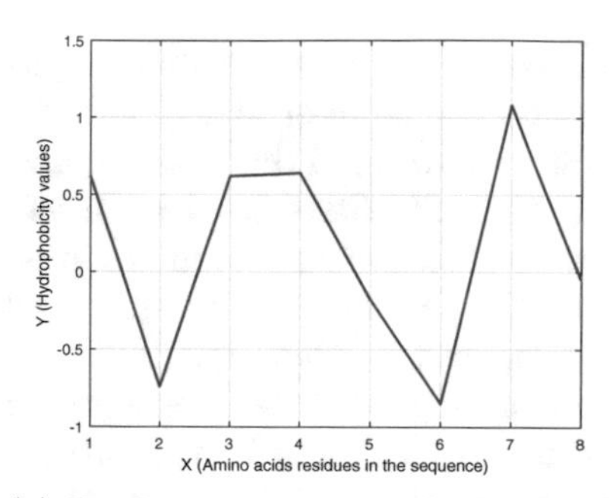

(a) Representation using hydrophobicity values

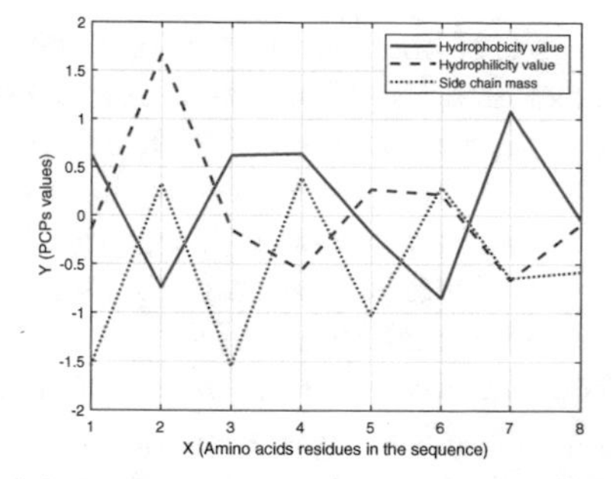

(b) Representation using three PCPs (hydrophobicity, hydrophilicity, side chain mass)

Fig. 1. Graphical representation of HIV-1 peptide 'AEAMSQVT' using different PCPs.

transformed into a sequence of moments. Each moment combines information from the whole curve. The moments are defined as

$$M_j = \sum_{i=1}^{n} \frac{(x_i - y_i)^j}{n^j} \qquad j = 1, 2, \ldots, n \tag{1}$$

where n is the length of the sequence and (x_i, y_i) are the coordinates of the ith point. Only the first n moments are independent, and higher moments do not add information. We collect the first (up to) n moments $(M_1, M_2, M_3, \ldots, M_n)$ in a moment vector. The correspondence between a protein sequences and its moment vector is one-to-one [21]. However, it is often useful to consider fewer moments, since the first moments are thought of as being most informative. When agreeing on a fixed number of moments, independent of the length of the sequence, then the moment vector representation solves the problem of mapping variable-length sequences to a fixed-length feature descriptor.

Each PCP describes an amino acid by one scalar value, hence each PCP can be used to represent the sequence graphically as a curve, as shown in Fig. 1b. The figure illustrates the graphical representation of a HIV-1 peptide using the standardized values of three PCPs: hydrophobicity, hydrophilicity, and side chain mass [6].

In previous work, multiple PCPs were aggregated into a single number [20, 22], resulting in a single moment vector per sequence. In this study we avoid such an aggregation step and use moments of the native PCP values to represent sequences. For example, in Fig. 1b each amino acid is described by three values, resulting in three curves for each sequence, and hence in three moment vectors. Each moment vector describes one PCP. In general when using P PCPs, we concatenate P moment vectors to form a feature vector for learning:

$$FeatureVector = \left[M_1^1, M_2^1, \ldots, M_n^1, M_1^2, M_2^2, \ldots, M_n^2, \ldots, M_1^P, M_2^P, \ldots, M_n^P\right] \tag{2}$$

The length of the feature vector depends on the length of the protein sequence (or the chosen length of the moment vector) and the number of selected PCPs.

To compare our results with previous work we have used different commonly used sets of PCPs to represent the protein sequences. These sets are: One PCP was used by Stephen et al. [21], three PCPs were used by Chou et al. [6] in order to test the PseAAC method, six PCPs were used by Sun et al. [20], and a set of 50 non-redundant PCPs of amino acids was proposed by Georgiev [9].

Concatenated moment vectors can yield rather high-dimensional representations, in particular when using large sets of up to 50 PCPs. In this situation, removing uninformative and redundant features can improve the performance of machine learning algorithms [2]. We tested two representative classes of dimensionality reduction techniques: *feature selection* based on the Random Forest (RF) algorithm [4], and *feature extraction* based on Principle Component Analysis (PCA) [10]. In both cases we tuned the target dimension (number of features) to the task at hand.

3 Support Vector Machines

In this study, we emphasized on studying the effect of using moment vector applied directly on PCPs for enhancing the performance of protein prediction. So to keep the work manageable, we restricted ourselves to using only the Support Vector Machine (SVM) to evaluate these new set of descriptors, since SVM has been recognized as one of the most powerful binary classifiers, and it is widely used in bioinformatics [12]. The main idea behind SVM is to find the optimal hyper-plane that separates the classes by maximizing the margin between the nearest data point of each class, called support vectors [7].

Linearly non-separable problems are approached by allowing for outliers (soft margin SVM) and by transforming the data into a high-dimensional feature space using a kernel function. Kernels turn SVMs into extremely flexible non-linear and even non-parametric classifiers. In this work we rely in the Gaussian Radial Basis Function (RBF) kernel, which is widely used and often results in good accuracy [16]. The SVM model with RBF kernel depends on two parameters: regularization ($C > 0$) and kernel parameter ($\gamma > 0$), both of which require problem-specific tuning.

In the preset study, SVM classifies were fed with moment vectors extracted from various sets of PCPs, as described in the previous section. The SVM hyper-parameters C and γ were tuned with grid search and 10-fold cross validation on each candidate feature set.

4 Benchmark Datasets

To evaluate the performance of the proposed method, we have used four datasets, two of which are peptide sequences (of length less than 50 [14]), and two of which are long sequences: Caspase 3 and human immunodeficiency virus type 1 (HIV-1) are peptide sequences, while Membrane proteins and DNA-binding proteins represent full protein sequences and large segment of proteins, respectively. We have split each dataset into a training and a testing set.

Caspase 3 plays a major role in programmed cell death as well as other vital cellular processes. Identification of Caspase 3 novel substrates is crucial to advance the understanding of the biological roles of this important enzyme. We have used the dataset of Caspase 3 human substrates [3], where all sequences have the same lengths of 14 amino acids. This dataset contains 247 mapped cleavage site, as well as 247 non-cleaved peptides extracted at random [3].

Membrane proteins plays an important role in genome sequencing. These proteins are embedded on one side of the cell membrane, either on the outer surface or the interior wall. The discrimination of the outer proteins from the inner is of medical importance [1]. In this study we used the dataset constructed by Park and coworkers [15]. The dataset contains 208 outer membrane proteins (OMPs) and 206 α-helical membrane proteins.

DNA-binding proteins play an important role in gene-regulation. Proteins target specific DNA sequences can be a potential cure for some of genetic diseases and cancers [11]. In this study we have used the DNAset dataset [11]. This dataset contains 145 DNA-binding and 226 non-binding proteins.

HIV-1 is an important enzyme in viral development and a causative agent of the acquired immune deficiency syndrome (AIDS) [19]. Predicting HIV-1 protease cleavage sites helps in finding and developing protease inhibitors, which can inhibit the replication of HIV-1 causing AIDS [19]. In the present study, we used a dataset from [17]. It contains 746 sequences: 401 represent cleaved 345 represent non-cleaved sites. All sequences have the same lengths of 8 amino acids.

5 Experimental Evaluation

In this section we present an experimental evaluation of our approach. We aim to answer the following questions:

- How does our approach of applying moment vectors directly to PCPs compare to alternative encoding methods?
- Which set of PCPs is best suited?
- Do we need dimensionality reduction, and if so, which method works best?

To evaluate the classification performance of our approach we rely on six common metrics: accuracy (ACC), sensitivity (SN), specificity (SP), precision (PRE), F-measure, and Matthew's Correlation Coefficient (MCC) [13]. In addition we considered the Receiver Operating Characteristics (ROC) curve, generated by varying the threshold of the binary classifier. This curve illustrates the trade-off between true positive rate (TPR) and false positive rate (FPR). The most common method for assessing the quality of the ROC curve is to compute the area under the ROC curve (AUC) [8]. When tuning parameters with cross-validation, we always relied on AUC. All reported numbers are computed on independent test sets.

5.1 Impact of the Underlying Set of PCPs

In order to study the effect of the proposed feature vector on the classification model, we run several experiments varying the set of PCPs. These experiments used the full feature vector to represent the protein sequence. As mentioned before, the proposed encoding consists of multiple moment vectors concatenated to form the full feature vector. The number of these moment vectors equal the number of selected PCPs. The length of each moment vector is restricted by the shortest sequence in the dataset. For full protein sequences this proceeding results in rather high-dimensional feature vectors. We restricted the length of each moment vector to a maximum of 15, since higher moments did not yield any improvement.

Table 1. Performance of four benchmarks, using the proposed feature vectors to encode the protein sequences varying the PCPs sets.

# of PCPs	Metric (%)						
	AUC	ACC	SN	SP	PRE	F-measure	MCC
Caspase 3 benchmark (the length of the shortest sequence is 14)							
1 PCP	52.03	53.28	55.74	50.82	53.12	54.40	06.57
3 PCPs	72.70	65.57	63.93	67.21	66.10	65.0	31.16
6 PCPs	70.14	63.93	62.30	65.57	64.41	63.33	27.88
50 PCPs	**74.33**	**69.67**	**67.21**	**72.13**	**70.69**	**68.91**	**39.39**
HIV-1 benchmark (the length of the shortest sequence is 8)							
1 PCP	80.16	73.66	76.0	70.93	75.25	75.62	46.97
3 PCPs	84.51	79.57	83.0	75.58	79.81	81.37	58.83
6 PCPs	87.90	82.80	85.0	80.23	83.33	84.16	65.35
50 PCPs	**96.79**	**90.86**	**92.0**	**89.53**	**91.09**	**91.54**	**81.61**
DNA binding benchmark (the length restricted to 15)							
1 PCP	82.44	69.57	83.33	60.71	57.69	68.18	43.36
3 PCPs	90.28	80.43	69.44	87.50	78.12	73.53	58.35
6 PCP	**91.72**	**81.52**	**72.22**	**87.50**	**78.79**	**75.36**	**60.77**
50 PCPs	81.25	77.17	52.78	92.86	82.61	64.41	51.43
Membrane proteins benchmark (the length restricted to 15)							
1 PCP	74.93	70.59	78.43	62.75	67.80	72.73	41.69
3 PCPs	85.81	79.41	88.24	70.59	75.0	81.08	59.76
6 PCPs	**88.35**	**86.27**	**94.12**	**78.43**	**81.36**	**87.27**	**73.46**
50 PCPs	79.85	72.55	80.39	64.71	76.74	74.55	45.66

Table 1 shows the performance metric values for the benchmark datasets using different selected sets of PCPs. We observe significant differences in performance.

For the peptide sequences Caspase3 and HIV-1 using 50 PCPs worked best. The same set of not a good choice for DNA binding and membrane protein datasets, where feature vectors based on six PCPs gave the best results. We conclude that large sets of PCPs work well overall, but problem-specific tuning is beneficial, since there is not a single best choice for all data sets.

5.2 Impact of Feature Selection

In Table 1, up to 700 features were used for classification. For dimensionality reduction we use two feature selection techniques. A simple baseline technique is to restrict each moment vector to its first n components. A more elaborate technique is to select moments using random forest based on Gini importance [4]. We tuned two hyperparameters of the random forest: the number of trees (values between 10 and 800), and the number of variables tried at each split (values around the square root of the total number of features). After applying the RF, we selected the initial set of moments from the feature vector using Gini importance values, then we used the forward selection technique to select the final subset of moments maximizing performance.

Table 1 illustrates that the best performance for the Caspase 3 and HIV-1 benchmarks is achieved with feature vectors based on 50 PCPs. On the other hand, the best performance for the other two datasets is achieved when we used the feature vectors based on 6 PCPs. For a direct comparison we have applied the feature selection techniques on those feature vectors.

The results are presented in Table 2. Major performance differences show up for the DNA binding and membrane protein datasets. RF-based selection is superior in 3 out of 4 cases, however, it fails on the membrane protein problem. In all cases, performance improved over using the full set of features. We conclude that dimensionality reduction is helpful, but the method of choice is problem-dependent.

Table 2. The classification performance using selection techniques to reduce the feature vectors for four benchmarks

Benchmarks	Best # of moments/PCs	Metric (%)						
		AUC	ACC	SN	SP	PRE	F-measure	MCC
Caspase 3	(1) The first 5 moments from each PCP	75.81	70.49	65.57	75.41	72.73	68.97	41.18
	(2) 420 selected moments using RF	75.19	71.31	72.13	70.49	70.97	71.54	42.63
HIV-1	(1) The first 7 moments from each PCP	96.91	91.40	92.0	90.7	92.0	92.0	82.70
	(2) 25 selected moments using RF	96.85	92.47	95.0	89.53	91.35	93.14	84.89
DNA-binding	(1) The first moment from each PCP	91.82	83.7	75.0	89.29	81.82	78.26	65.41
	(2) 20 selected moments using RF	89.20	86.27	100	72.55	78.46	87.93	75.45
Membrane proteins	(1) The first moment from each PCP	93.04	90.2	98.04	82.35	84.75	90.91	81.40
	(2) 35 selected moments using RF	88.77	87.25	98.04	76.047	80.65	88.50	76.31

We have noticed that in most cases and for different PCPs sets we need only the first few moments from each PCP to represent long protein sequences. On the other hand, for peptide sequences we need more than the first moments to achieve good performance. Our results indicate that for peptide sequences a number of moments around ($Sequence\ Length/2$) is a good general recommendation.

5.3 Impact of Feature Extraction

As an alternative to selection a subset of a large number of given features, here we use PCA to construct new (linearly transformed) features. The only hyperparameter of the method is the number of principal components (PCs). We tested different automated approaches [5], but it turns out that there is no ideal method we can depend on for all cases in this study. Therefore we tuned the parameter with cross validation for optimal performance. We found that for good results, the selected PCs should cover at least 98% of the total variance, i.e., only up to 2% of the information is discarded.

Table 3. The classification performance using PCA algorithm to reduce the feature vector contains moments applied to different PCPs.

Benchmark	PCPs sets	# of moments used	Best# of PCs	Metric (%)						
				AUC	ACC	SN	SP	PRE	F-measure	MCC
Caspase 3	3 PCPs	All moments (14)	19	82.05	77.87	73.77	81.97	80.36	76.92	55.93
HIV-1	50 PCPs	All moments (8)	120	97.13	90.32	86.0	95.35	95.56	90.53	81.16
DNA-binding	50 PCPs	8 moments	42	92.16	89.13	88.89	89.29	84.21	86.49	77.49
Membrane proteins	3 PCPs	All moments (15)	40	99.12	97.06	100	94.12	94.44	97.14	94.28

We applied the PCA with whitening to two different feature vectors using the four different PCPs sets. These feature vectors are the whole feature vector that contains the all moment vectors applied to four different PCP sets, and the vector containing the selected equal number of moments applied to each PCP that achieved best performance in Table 1. The best results of classification performance using PCA is found in Table 3.

It is evident that in most cases whitened PCA features are superior to selected feature selection. A notable exception is the HIV-1 dataset, where 25 features selected by a random forest yield better results (cf. Table 2).

5.4 Comparison with the State of the Art

To verify the performance of the proposed encoding method, we compared our test results with other encoding methods based on moment vectors. Furthermore, we compared the results of our proposed encoding with one of the most widely used encoding methods, namely PseAAC. For details of this method we refer to [6].

Table 4. Comparison of the proposed approach with previous moment-based encoding methods and PseAAC.

Encoding Method	Metric (%)						
	AUC	ACC	SN	SP	PRE	F-measure	MCC
Caspase 3 benchmark							
(1) moment vector based on graph energy, best results with 2 moments	54.45	54.92	52.46	57.38	55.17	53.78	09.85
(2) moment vector based on scaled hydrophobicity, best results with 3 moments	54.35	51.06	59.99	45.49	49.18	50.87	05.50
(3) PseAAC with 50 PCPs	73.04	68.85	60.66	77.05	72.55	66.07	38.22
(4) our approach, best results with 19 PCs, from full moment vectors of 3 PCPs	**82.05**	**77.87**	**73.77**	**81.97**	**80.36**	**76.92**	**55.93**
HIV-1 benchmark							
(1) moment vector based on graph energy, best results with 7 moments	67.44	68.28	82.0	52.33	66.67	73.54	36.16
(2) moment vector based on scaled hydrophobicity, best results with 4 moments	78.09	75.27	86.0	62.79	72.88	78.9	50.51
(3) PseAAC with 1 PCP	96.91	90.32	94.0	86.05	88.68	91.26	80.61
(4) our approach, best results with 25 moments, moments selected by RF algorithm	**96.85**	**92.47**	**95.0**	**89.53**	**91.35**	**93.14**	**84.89**
DNA-binding benchmark							
(1) moment vector based on graph energy, best results with 2 moments	82.59	72.83	83.33	66.07	61.22	70.59	48.33
(2) moment vector based on scaled hydrophobicity, best results with 4 moments	84.13	79.35	69.44	85.71	75.76	72.46	56.13
(3) PseAAC with 3 PCPs	94.0	89.13	86.11	91.07	86.11	86.11	77.18
(4) our approach, best results with 42 PCs, from first 8 moments of 50 PCPs	**92.16**	**89.13**	**88.89**	**89.29**	**84.21**	**86.49**	**77.49**
Membrane proteins benchmark							
(1) moment vector based on graph energy, best results with 8 moments	68.47	63.73	52.94	74.51	67.5	59.34	28.11
(2) moment vector based on scaled hydrophobicity, best results with 9 moments	94.93	91.18	98.04	84.31	86.21	91.74	83.14
(3) PseAAC with 3 PCPs	98.42	95.10	98.04	92.16	92.59	95.24	90.35
(4) our approach, best results with 40 PCs, PCA on full moment vectors from 3 PCPs	**99.12**	**97.06**	**100**	**94.12**	**94.44**	**97.14**	**94.28**

The two main previous methods used the concept of moment vector to encode the protein sequences are:

- Yau et al. [21] use a scaled descriptor of the hydrophobicity values to compute the moment vectors for sequences. Compared to our method, it relies on a single PCP and introduces an additional rank-based rescaling step.
- Sun et al. [20] represent the relationships between amino acids as a graph by using the values of six PCPs. They reduce the information to a single number of amino acid by aggregating the graph properties into an energy function. Each protein sequence is represented by its moment vector based on the graph energy.

Both methods were tested on clustering problems. In this study we applied them to four supervised classification problems. We compared our results to the existing approaches by training SVMs on the same data sets, but with all sequences encoded by the feature sets proposed in [20,21].

Table 4 lists the performance of our encoding and the previous approaches side-by-side. This table contains the best performance we can achieve for each benchmark based on different PCP sets and numbers of moments. The results illustrate the superiority of our encoding approach over previous approaches across all four benchmarks, and across all performance metrics. We also notice that the encoding of Yau et al. is still better than that of Sun et al. for most benchmarks except for Caspase-3, where the performance of both approaches is poor.

6 Conclusion

This study has evaluated a new approach for using moment-based representations to encode the protein sequences. We applied moments directly to the native PCPs values to solve different prediction problems. We have systematically varied the set of PCPs, some of which result in high-dimensional feature vectors. Therefore we have also compared different approaches to dimensionality reduction, namely feature selection with random forests and feature extraction with principal component analysis.

The results show that the proposed approach of encoding can enhance the performance of the prediction for four protein benchmarks compared to the previous approaches and the Pseudo Amoni Acid Composition (PseAAC) method. This enhancement depends on the selected PCPs, number of moments used, and the type of the problem. One the other hand, our results show that this encoding approach can generate highly correlated features. Decorrelating these features using PCA can be significantly enhance predictive performance on most benchmarks. Furthermore, the results demonstrate that our encoding approach is applicable for both peptide and full protein sequences.

References

1. Almen, M., Nordström, K., Fredriksson, R., Schioth, H.: Mapping the human membrane proteome: a majority of the human membrane proteins can be classified according to function and evolutionary origin. BMC Biol. (2009)

2. Alpaydın, E.: Introduction to Machine Learning. The Adaptive Computation and Machine Learning Series, 2nd edn. Massachusetts Institute of Technology (2010)
3. Ayyash, M., Tamimi, H., Ashhab, Y.: Developing a powerful in Silico tool for the discovery of novel caspase-3 substrates: a preliminary screening of the human proteome. BMC Bioinf. (2012)
4. Breiman, L.: Random forests. Mach. Learn. **45**(1), 5–32 (2001)
5. Cangelosi, R., Goriely, A.: Component retention in principal component analysis with application to cDNA microarray data. Biol. Dir. **2**(2) (2007)
6. Chou, C.: Prediction of protein cellular attributes using pseudo-amino-acid composition. In: PROTEINS: Structure, Function, and Genetic, pp. 246–255 (2001)
7. Cortes, C., Vapnik, V.: Support-vector networks. Mach. Learn. **20**(3), 273–297 (1995)
8. Fawcett, T.: An introduction to ROC analysis. Pattern Recogn. Lett. **27**, 861–874 (2006)
9. Georgiev, A.: Interpretable numerical descriptors of amino acid space. J. Comput. Biol. **16**(5) (2009)
10. Jolliffe, I.: Principal Component Analysis, 2nd edn. Springer, New York (2002)
11. Kumar, M., Gromiha, M.M., Raghava, G.P.S.: Identification of DNA-binding proteins using support vector machines and evolutionary profiles. BMC Bioinf. **8** (2007)
12. Liu, B., Xu, J., Lan, X., Xu, R., Zhou, J., Wang, X., Chou, K.C.: iDNA-Prot—dis: identifying DNA-binding proteins by incorporating amino acid distance-pairs and reduced alphabet profile into the general pseudo amino acid composition. PLoS ONE **9** (2014)
13. Matthews, B.W.: Comparison of the predicted and observed secondary structure of T4 phage lysozyme. Biochimica et Biophysica Acta (BBA) - Protein Structure **405**(2), 442–451 (1975)
14. McKee, M., McKee, J.: Biochemistry: The Molecular Basis of Life, 5th edn. Oxford University Press, Oxford (2011)
15. Park, K., Gromiha, M., Horton, P., Suwa, M.: Discrimination of outer membrane proteins using support vector machines. Bioinformatics **21**, 223–229 (2005)
16. Qu, K., Han, K., Wu, S., Wang, G., Wei, L.: Identification of DNA-binding proteins using mixed feature representation methods. Molecules **10** (2017)
17. Rognvaldsson, T., You, L., Garwicz, D.: State of the art prediction of HIV-1 protease cleavage sites. Bioinformatics **31** (2015)
18. Saidi, R., Maddouri, M., Nguifo, E.: Protein sequences classification by means of feature extraction with substitution matrices. BMC Bioinf. (2010)
19. Singh, O., Chia-Yu, E.: Prediction of HIV-1 protease cleavage site using a combination of sequence, structural, and physicochemical features. BMC Bioinf. **17** (2016)
20. Sun, D., Xu, C., Zhang, Y.: A novel method of 2D graphical representation for proteins and its application. Commun. Math. Comput. Chem. **75**, 431–446 (2016)
21. Yau, S.S.T., Yu, C., He, R.: A protein map and its application. DNA Cell Biol. **27** (2008)
22. Zhou, X., Li, X., Li, M., Lu, X.: Predicting protein functional class with the weighted segmented pseudo-amino acid composition moment vector. Commun. Math. Comput. Chem. **66**, 445–462 (2011)

A Hybrid of Particle Swarm Optimization and Minimization of Metabolic Adjustment for Ethanol Production of *Escherichia Coli*

Mee K. Lee[1], Mohd Saberi Mohamad[2,3(✉)], Yee Wen Choon[1], Kauthar Mohd Daud[1], Nurul Athirah Nasarudin[1], Mohd Arfian Ismail[4], Zuwairie Ibrahim[5], Suhaimi Napis[6], and Richard O. Sinnott[7]

[1] Artificial Intelligence and Bioinformatics Research Group, School of Computing, Faculty of Engineering, Universiti Teknologi Malaysia, 81310 Skudai, Johor, Malaysia
meekhee91@gmail.com, ewenchoon@gmail.com, kautharmohdaud@yahoo.com, nathirah56@live.utm.my

[2] Institute for Artificial Intelligence and Big Data, Universiti Malaysia Kelantan, City Campus, Pengkalan Chepa, 16100 Kota Bharu, Kelantan, Malaysia
saberi@umk.edu.my

[3] Faculty of Bioengineering and Technology, Universiti Malaysia Kelantan, Jeli Campus, Lock Bag 100, 17600 Jeli, Kelantan, Malaysia

[4] Soft Computing and Intelligent System Research Group, Faculty of Computer Systems and Software Engineering, Universiti Malaysia Pahang, 26300 Kuantan, Pahang, Malaysia
arfian@um.edu.my

[5] Faculty of Manufacturing Engineering, Universiti Malaysia Pahang, Pekan, Pahang, Malaysia
zuwairie@ump.edu.my

[6] Faculty of Biotechnology and Biomolecular Sciences, Universiti Putra Malaysia, 43400 UPM Serdang, Selangor, Malaysia
suhaimi@upm.edu.my

[7] School of Computing and Information Systems, The University of Melbourne, Melbourne, VIC 3052, Australia
rsinnott@unimelb.edu.au

Abstract. Ethanol is a chemical-colourless compound that widely used in pharmaceutical, medicines, food products, and industrial applications. As the demand for ethanol is rising recently, attention has been given on metabolic engineering of Escherichia coli (*E.coli*) to enhance its production through alteration of its genetic content. This research mainly aimed to optimize ethanol production in *E.coli* using a gene knockout strategy. Several gene knockout strategies like OptKnock and OptGene have been proposed previously. However, most of them suffer from premature convergence. Hence, a hybrid of Particle Swarm Optimization (PSO) and Minimization of Metabolic Adjustment (MOMA) algorithm is proposed to identify the list of gene knockouts in maximizing the ethanol production and growth rate of *E.coli*. Experiment results show that the hybrid method is comparable with two state-of-the-art methods in term of growth rate and production.

F. Fdez-Riverola et al. (Eds.): PACBB 2019, AISC 1005, pp. 36–44, 2020.
https://doi.org/10.1007/978-3-030-23873-5_5

Keywords: Particle swarm optimization · Minimization of metabolic adjustment · Metabolic engineering · Bioinformatics · Artificial intelligence

1 Introduction

The advance of biotechnology helps to make a conversion of biomass into fuel ethanol with low-cost productivity. Although biomass has become an excellent resource for the conversion of sugar into ethanol; however, the amount produced by these microbes are far below the maximum industry threshold [1]. Hence, great attention has been given on metabolic engineering of microorganisms namely *E.coli*, to optimize the production of ethanol. Gene knockout involved the inactivation of a particular gene responsible for a significant reaction such as increased flux in ethanol production. Different number of knockouts genes result in different reaction rate. Several methods are available for the study of gene knockouts in microbes. The methods can be classified according to their mathematical formulations and optimization frameworks, namely metaheuristics, bi-level mixed integer programming methods and elementary mode analysis. In this paper, we only focus on metaheuristics based methods such as OptKnock, OptGene, and set-based evolutionary algorithms (SEAs) and simulated annealing (SA) algorithms. OptKnock is the first method that becomes the template for developments of other constraint-based strain design methods [2]. OptKnock associated two competing objectives functions that are optimized simultaneously. The bi-level formulation of OptKnock can be transformed into single optimization by accumulating the objective functions.

Meanwhile, different heuristic methods have been applied although it does not guarantee to find an optimal solution, yet it founds near-optimal solution and computationally less expensive. OptGene is the first method that applies heuristic method, genetic algorithm (GA) to identify reactions/genes knockout for overproduction of metabolites. OptGene solves a problem by searching the highest number gene knockouts and the fitness is evaluated using Flux Balance Analysis (FBA). SA and SEA have been introduced by Rocha and co-workers that reformulate the outer optimization problem of OptGene by allowing more compact representation of number of knockouts. SEA and SA allow for binary and integer representations of reactions knockout.

Unfortunately, the three methods have their limitations. OptKnock provides overly optimistic solutions as it selects the best solution with the highest production yield despite the minimum growth rate. OptGene has a high possibility of failure to find the best solution as the solutions may be trapped in the local optimal search. Moreover, GA used by OptGene suffers high computation time [3]. Hence, a hybrid algorithm of PSO and MOMA (PSOMOMA) is proposed to overcome the limitation as mentioned above. PSO looks for the best solution to a problem, whereas MOMA is used to evaluate the fitness of a solution found by calculating ethanol production. PSOMOMA was designed to identify the best number of gene knockouts that result in optimizing the production of ethanol and growth rate.

2 Materials and Methods

2.1 Dataset and Experiment Setup

An updated version of the iAF1260 *E.coli* model named IJO1366 [4] was being used. IJO1366 is a metabolic network reconstruction model of *E.coli* that in SBML format, which is available in BioModels Database, KEGG. Table 1 below shows the pre-processed results of the model. There are two built-in functions in COBRA toolbox used during the model pre-processed; removeDeadEnds and reduceModel. In removeDeadEnds function, the model is scanned through to determine which metabolites are either can only be consumed or produced. This can be seen in their respective lower bound values whether zero (0) or not. In the latter function, reduceModel, the reactions that never used are removed from the model. The model is scanned through and reactions that below the threshold are removed.

Table 1. The result of pre-processed model of iJO1366 *E.coli*

Model	Number of reaction	Number of metabolites
iJO1366 Model	2583	1805
Pre-processed model	2342	1585
Removed Amounts	241	220

Table 2 shows the parameter setting for PSOMOMA. These values are determined during the pre-experimental setting. We have tested 20–100 for a number of iterations and allowed PSOMOMA to run until 100 runs. Referring to the correction factor and inertia, these values are determined previously [5, 6]. The experiment was implemented in MATLAB R2010b with COBRA toolbox and SBML toolbox associated to it. The source code for the method is available upon request.

Table 2. Parameter settings for PSOMOMA

Parameter	Value
Number of iterations	20
Inertia	1.0
Correction factor	2.0
Number of particles	2500
Number of runs	50
Gene knockouts (every 50 runs)	2, 3, 4, 5

2.2 PSO Algorithm

PSO is a heuristic optimization algorithm and initially introduced by Kennedy in 1995 based on the movement behaviour of bird flocks or fish schools [7].

2.3 MOMA

MOMA is used for fitness calculation of mutants flux states [10]. Whenever perturbation occurs in a metabolic network, the behaviour of this perturbed network can be simulated by MOMA since it can calculate a solution with minimum changes. In other words, MOMA helps to study how well a cell can adapt itself in term of their biochemical production and growth rate after gene knockouts occur in the mutant strain.

2.4 A Hybrid of PSO and MOMA (PSOMOMA)

MOMA is able to simulate the behaviour for a perturbed metabolic network, but, it could not find an optimized solution of gene knockouts. Hence, PSOMOMA is introduced to maximize the production of ethanol and growth rate by performing a different number of gene knockouts. MOMA is used in this hybrid algorithm to evaluate and determine whether the gene knockout causes any lethal or inability for optimized ethanol production and growth rate. The flowchart for PSOMOMA is shown in Fig. 1 and processes involved are describe below.

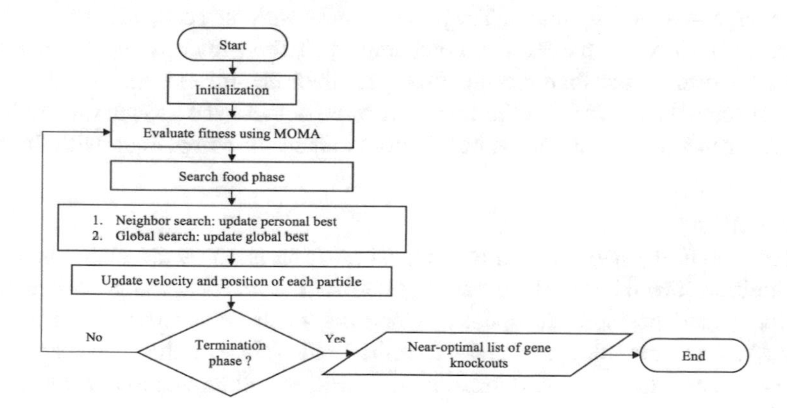

Fig. 1. Flowchart of PSOMOMA.

Initialization

PSOMOMA starts by initializing parameter settings and generating particles. Particle positions are randomly given a value of 0 or 1, where the 0 value indicates the gene cannot be knockout whereas 1 otherwise. Figure 2 shows the representation of particle positions, whereby each gene is account for a single enzyme used in a reaction.

Fitness Evaluation

After generating the particle, fitness of the genes knockouts was calculated using MOMA. Particles with the best fitness are stored. Particles with a growth rate greater or equal to 0.1 hr^{-1} are only to be considered.

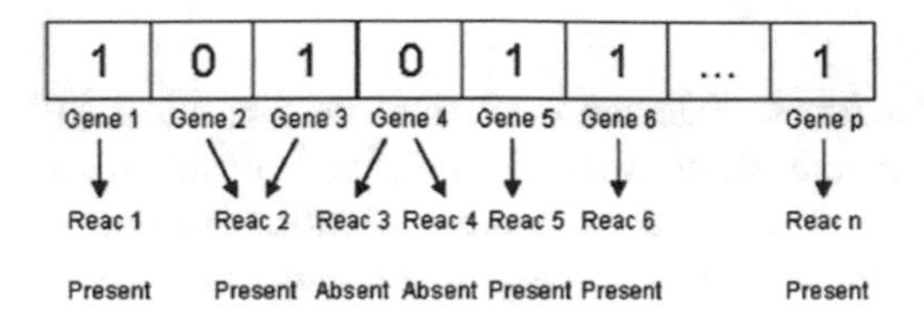

Fig. 2. The representation of particle positions

Search Food
This phase involves the initialization of both the velocity and position of each particle. Initially, their velocity was set to the value of 0. The position of particles represents the bird flocks or fish swarms searching foods in a different location that indicates the ethanol product obtained from each population.

Neighbour Search
Neighbour search phase applies the idea ring topology whereby, the information disseminates slowly thus when one particle has a rise in its fitness, the closely particles will converge to a local optimum. The process starts with the second iteration whereby, the current position of particles is compared with the previous one. If the second position performs better than the previous one, then the current position is set a personal best (pbest) provided that the current particle possessed a growth rate higher than 0.01. As for the next iteration, the pbest only is taken for comparison with the current position.

Global Search
In global search, the particle with the highest position is set as the global best (gbest) which indicates that it has the best amount of ethanol production and growth rate. Then, the velocity and position are updated according to the gbest particle. This can be visualized as a star topology, whereby a central node (gbest) influenced other particles that are connected to it. For each iteration, all particles will move towards the gbest and adjust the position accordingly.

Termination
Searching for particular pbest and gbest is continued iteratively until some 20 times have been reached. The position and velocity obtained from both pbest and gbest show that they are the fittest among the population and hence result in best productivity of ethanol and growth rate.

3 Result and Discussion

3.1 Experimental Results

Table 3 shows the best result obtained from the PSOMOMA among 50 runs of respective gene knockouts numbers. We compare the results of our method with previous works and validate the functions of knockout genes based on literature. The glucose uptake rate is set at 10 mmolgDW^{-1}hr^{-1} and used as the sole carbon source.

Table 3. Ethanol production and growth rate of *E.coli* obtained from 50 runs of PSOMOMA

No. of genes knockout	Knockout list	Ethanol production (mmolgDW^{-1}hr^{-1})	Growth rate (hr^{-1})
2	ACKr, PPS	17.2029	0.1892
3	pflA, frdB, ldhA	**17.2270**	0.1876
4	ACKr, ldhA, FUMt2_2, fdhF	16.4891	0.1944
5	ACKr, fumB, PPS, GND, GLUDy	16.4501	**0.2338**

Note: Bold style represents the best result

The first set of knockouts were *ACKr* and *PPS*, which encoded by genes *ackA* and *ppsA*, respectively. The result shows that *E.coli* is able to survive with a growth rate of 0.1892. Besides, the production rate of ethanol was the second highest among the other number of knockouts, which is 17.2029. As stated by [8], the conversion of acetyl-phosphate to acetate was inhibited with the elimination of the *ackA*. Hence, acetyl Co-A now can directly convert into ethanol without reconverting back into acetate. Besides, [9] stated that PPS was a phosphoenolpyruvate (PEP) synthase that catalyzes the phosphorylation of pyruvate to PEP. Hence, knockout of PPS causes the accumulation of pyruvate which then accelerate ethanol production.

Another set with three gene knockouts (*pflA, frdB,* and *ldhA* genes) yields the highest production of ethanol. NADH was the main substrate used for the production of lactate, succinate, acetate, and ethanol. According to [10], *the pflA* gene is for the production of acetate, whereas *frdB* and *ldhA* were responsible for the production of succinate and lactate. By removing these three genes, it causes a flux in ethanol production, which is 17.2270, with a growth rate of 0.1876.

The third set of gene knockouts involves the knockout of *ACKr, ldhA, FUMt2_2,* and *fdhF*, which yields a growth rate of 0.1944. *ldhA* and *ACKr* increase ethanol production. Knockout of *FUMt2_2* causes poor aerobic growth on succinate, which induced the increase in ethanol due to the lack of succinate competitor for the NADH substrate of ethanol production. *fdhF* gene was also responsible for the regulation of formate in *E.coli*. Deletion of the *fdhF* gene causes reconversion of formate back into pyruvate. An increased level of pyruvate is then converted into ethanol. Hence, produced 16.4891 of ethanol.

The last set of gene knockout is *ACKr, fumB, PPS, GND,* and *GLUDy* genes. Glucose-6-phosphate level increases with the knockout of GND. Hence, there is more substrate for increasing the production of ethanol. Knockout of *fumB* causes the inhibition of the conversion of malate into fumarate, thus no substrate (fumarate) for succinate production, where this indicates there is a lack of competitor for *NADH* in ethanol production. Also, *GLUDy* was also a necessity for the conversion of isocitrate to α-ketoglutarate. Therefore, there is no more α-ketoglutarate for the conversion into succinate, which induced ethanol production. Moreover, knockout of *ACKr* and *PPS* also help to increase the production of ethanol, which yields 16.4501 with a growth rate of 0.2338. As a conclusion, three number of gene knockouts (*pflA, frdB, ldhA*) show the highest production of ethanol, which yields 17.2270. While, five number of gene knockouts (*ACKr, fumB, PPS, GND, GLUDy*) show the highest growth rate of 0.2338.

3.2 Comparative Analysis

The sets of gene knockouts are validated according to their contribution in optimizing the ethanol production and growth rate of *E.coli* (refer Figs. 3 and 4). Besides, the growth rate and the production values are compared with previous works to prove that this proposed method yields better performance than the earlier methods. In this research, we compared the PSOMOMA result obtained from different sets of gene knockouts with the knockout result from OptFlux [10] and OptReg [11].

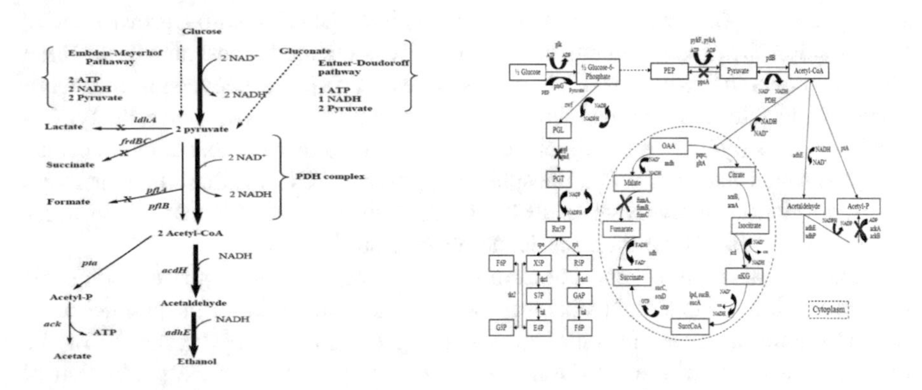

Fig. 3. Ethanol production pathway in *E.-coli* for knockout of *pflA, frdB,* and *ldhA* genes [10].

Fig. 4. Ethanol production pathway in *E.coli* for knockout of *ACKr, fumB, PPS, GND,* and *GLUDy* genes.

In Table 4, PSOMOMA performs better for two genes knockout compared to OptFlux. The growth rate is 0.1892, which is slightly higher than OptFlux. In term of production, PSOMOMA is better than OptReg. As for the growth rate of *E.coli*, mutant with three gene knockouts in PSOMOMA performs better than OptReg. This is because PSOMOMA provides a greater differential in solving an unconstrained non-linear problem with continuous design variables, whereas OptFlux and OptReg provide a smaller range of differential; eventually drive a higher production of ethanol using PSOMOMA. Hence, PSOMOMA produces an optimize ethanol production with two gene knockouts and higher growth rate with three gene knockout mutant of *E.coli* as compared to OptReg and OptFlux method.

Table 4. Comparison of PSOMOMA, OptFlux, and OptReg

Method	No. of genes knockout	Gene knockout	Ethanol production ($mmolgDW^{-1}hr^{-1}$)	Growth rate (hr^{-1})
PSOMOMA	2	*ACKr, PPS*	**17.2029**	0.1892
	3	*pflA, frdB, ldhA*	17.2270	0.1876
	4	*ACKr, ldhA, FUMt2_2, fdhF*	16.4891	0.1944
	5	*ACKr, fumB, PPS, GND, GLUDy*	16.4501	**0.2338**
OptFlux [10]	1	*ΔldhA*	8.4851	0.1045
	2	*ΔpflA ΔldhA*	8.5106	0.1846
OptReg [11]	2	*O2t, PTAr.*	16.3000	0.1900
	3	*PGI, PFL, O2t.*	**18.7400**	0. 0800

Note: Bold style represents the best result.

4 Conclusion and Future Work

PSOMOMA is a reliable method as it utilizes MOMA to understand the flux in production and growth rate of *E.coli* by reducing the distance in flux space and performs fitness calculation for every set of gene knockouts. Generally, PSOMOMA performs better in maximizing the ethanol production and growth rate of *E.coli* as compared to OptFlux and OptReg methods. In future research, ROOM can be used with the PSO to enhance the metabolic rate of *E.coli*.

Acknowledgement. We would like to thank the Ministry of Education Malaysia for supporting this research by the Fundamental Research Grant Schemes (grant number: RDU190113 and R.J130000.7828.4F720).

References

1. Tang, P., Choon, Y.W., Mohamad, M.S., Deris, S., Napis, S.: Optimising the production of succinate and lactate in Escherichia coli using a hybrid of artificial bee colony algorithm and minimisation of metabolic adjustment. J. Biosci. Bioeng. **119**(3), 363–368 (2015)
2. Burgard, A.P., Pharkya, P., Maranas, C.D.: OptKnock: a bilevel programming framework for identifying gene knockout strategies for microbial strain optimization. Biotechnol. Bioeng. **84**(6), 647–657 (2003)
3. Arif, M.A., Mohamad, M.S., Abd Latif, M.S., Deris, S., Remli, M.A., Daud, M.K., Ibrahim, Z., Omatu, S., Corchado, J.M.: A hybrid of Cuckoo Search and Minimization of Metabolic Adjustment to optimize metabolites production in genome-scale models. Comput. Biol. Med. **102**, 112–119 (2018)
4. Orth, J.D., Conrad, T.M., Na, J., Lerman, J.A., Nam, H., Feist, A.M., Palsson, B.Ø.: A comprehensive genome-scale reconstruction of Escherichia coli metabolism. Mol. Syst. Biol. **7**(1), 535 (2011)

5. Klein, H.A., Shulla, A., Reimann, A.S., Keating, H.D., Wolfe, J.A.: The intracellular concentration of acetyl phosphate in escherichia coli is sufficient for direct phosphorylation of two-component response regulators. J. Bacteriol. **189**(15), 5574–5581 (2007)
6. Zhou, L., Zuo, R.Z., Chen, Z.X., Niu, D.D., Tian, M.K.: Evaluation of genetic manipulation strategies on D-lactate production by Escherichia coli. Curr. Microbiol. **62**(3), 981–989 (2011)
7. Kennedy, J., Eberhart, R.: Particle swarm optimization. In: Proceeding of the 1995 IEEE on Neural Networks, pp. 1942–1948 (1995)
8. Segre, D., Vltkup, D., Church, M.G.: Analysis of optimality in natural and perturbed metabolic networks. Proc. Natl. Acad. Sci. **99**(23), 15112–15117 (2002)
9. Mienda, S.B., Shamsir, S.M., Shehu, I., Deba, A.A., Galadima, A.I.: In silico metabolic engineering interventions of Escherichia coli for enhanced ethanol production, based on gene knockout simulation. J. Multi. Sci. Technol. **5**(2), 16–23 (2014)
10. Dien, S.B., Cotta, A.M., Jeffries, W.T.: Bacteria engineered for fuel ethanol production: current status. Appl. Microbiol. Biotechnol. **63**(3), 258–266 (2003)
11. Pharkya, P., Maranas, C.D.: An optimization framework for identifying reaction activation/inhibition or elimination candidates for overproduction in microbial systems. Metab. Eng. **8**(1), 1–13 (2006)

Cache-Efficient FM-Index Variants for Mapping of DNA Sequences

Jozef Sitarčík(✉) and Mária Lucká

Faculty of Informatics and Information Technologies,
Slovak University of Technology, Bratislava, Slovakia
{xsitarcik,maria.lucka}@stuba.sk

Abstract. Mapping reads to a reference genome is the first and very important step in genome analysis. One of the most used data structure for DNA read mapping is the FM-index. To this day many different variants of the FM-index exist. In the proposed paper we introduce two new variants of the FM-index suitable especially for DNA sequences. Proposed variants decrease the number of cache misses by efficiently interleaving auxiliary data structures of the FM-index thus increasing the searching speed. Experimental results have shown that the proposed variants are about two times faster than other variants, while having comparatively low memory requirements.
Source code is available at https://github.com/xsitarcik/DNASeqMap.

Keywords: FM-index · Read mapping · DNA sequences · Wavelet tree · Cache performance

1 Introduction

Sequencing large amount of genomes results in a tremendous amount of data that need to be processed. This poses a great challenge in managing the data effectively. The initial and key step in genome analysis is read mapping. One of the most used approaches to read mapping is the *seed-and-extend* approach [2], where seeds represent subsequences of input pattern P. Seeds are exactly mapped to a reference sequence T and then matched seeds are extended. To provide memory and time efficient exact mapping, the reference sequence is processed using the FM-index.

The FM-index was introduced in [6] and it is still very fast and efficient when searching short patterns [4]. The FM-index is based on the Burrows-Wheeler Transform (BWT) [5]. The FM-index utilizes the relationship between suffix array [19] and the BWT, where the BWT is used for suffix array compression [1,9]. In most implementations, BWT is stored as a wavelet tree - succinct data structure, first introduced in [14] and more in details described in [7,15,20]. The basic binary wavelet tree encodes the input sequence in bitvectors. A very important operation on the bitvectors is the $rank(pos)$ operation, that counts

F. Fdez-Riverola et al. (Eds.): PACBB 2019, AISC 1005, pp. 45–52, 2020.
https://doi.org/10.1007/978-3-030-23873-5_6

set bits up to the position *pos* [12]. A rank support structure is then used to store the precomputed ranks (number of set bits) to speed up this operation.

There are two important operations of the FM-index used in read mapping: *count*(P) and *locate*(P). *Count*(P) operation returns the number of occurrences of P in T, where P is an input pattern and T represents the reference sequence. *Locate*(P) operation returns all of the positions of occurrences of P in T. This is not trivial, because T is transformed by the BWT and the original string is thrown away, because the BWT is reversible. To speed up this operation, the suffix array is used. To decrease its high memory requirements, the suffix array is sampled and the non-stored values are computed on-the-fly by the *LF-mapping* operation of the FM-index [9].

Many different variants of the FM-index exist and are reviewed in [8,11]. Most of them are already implemented in the Succinct Data Structure Library (SDSL)[1], which was introduced in [10]. The most of the variants focus on decreasing memory complexity for example with usage of RRR-vectors [21] or by using run-length encoding [17,18]. Recently proposed variants also focus on compression improvements [16] or alphabet independence of the FM-index [3]. The big problem of the FM-index however are pseudo-random memory accesses during calculations of the *count*(P) or *locate*(P) operations, because the accessed values are spread pseudo-randomly on various positions due to the used transformation (BWT). These random accesses then cause many cache misses and slow down running time of both operations.

In this paper we present new cache-efficient variants of the FM-index focused on DNA sequences and read mapping. We also analyze the usage of a precomputed k-mer table for operation *count*(P) in the FM-index. We conduct experiments on two different datasets and compare proposed variants with other available versions of the FM-index.

2 Reducing Cache Miss Rate

The main idea of the cache miss rate reduction is to create a data structure denoted here as *ilrank*, which interleaves the precomputed ranks with the bitvector data [11,22]. So *ilrank* constructed for a bitvector with N bits is divided into n chunks $ch_1, ch_2, ...ch_n$, where the size of a chunk is s_{ch} and $n = \frac{N}{s_{ch}}$. This way cache miss rate is reduced, because precomputed ranks are closer to the corresponding data. However, to guarantee only one cache miss at most for calculating *rank*(*pos*), it is also necessary that the chunk size of *ilrank* is a multiple of the size of the cache line [13].

For the reference sequence T from an alphabet size σ it is necessary to calculate $lg(\sigma)$ rank operations on each level of the wavelet tree. This can cause up to $lg(\sigma)$ cache misses, however our proposed variants cause only one cache miss at most.

[1] https://github.com/simongog/sdsl-lite.

2.1 Proposed Variants

DNA sequences are an important application of the FM-index, hence our proposed variants are focused specifically on the DNA sequences, where alphabet $\Sigma = \{A, C, G, T\}$. In this case, the basic wavelet tree would consist of 3 bitvectors - root, left and right child, which we denote as *root*, *lc* and *rc*, respectively. To exploit this low number of bitvectors, we further build on the previously mentioned idea to reduce the cache miss rate by interleaving the data of all bitvectors with the corresponding ranks. Our new idea is to build a structure divided into chunks similarly as *ilrank*, where each chunk contains data from each of the bitvectors *root*, *lc* and *rc*, and also corresponding precomputed ranks for each bitvector. We propose two variants of the FM-index *wt_dna_il* and *wt_dna_il_sa* that implement this idea.

Both variants follow similar construction algorithm, which for a chunk ch_i is as follows:

1. load next 256 bits from *root* and store them as eight 4-byte words $a_0, a_1 ... a_7$.
2. load next X bits from *lc*, where X is the number of zero bits in $a_0, a_1 ... a_7$ and store them as $a_8 ... a_{8+X/32}$.
3. load $256 - X$ bits from *rc* and store them as 4-byte words in $a_{8+X/32} ... a_{15}$ and the right-to-left order, so the least significant bit corresponds to the first bit of that word and a_{15} corresponds to the first 4-byte word of *rc*. So we ensure that there is no need for calculating the value of X.
4. store the number of occurrences of each character A, C, G and T up to the position $i \times 256$ in the BWT on positions $a_{16} ... a_{19}$.
5. store precomputed ranks of bitvector *root* in 64 bits on positions a_{20}, a_{21}, where j-th 8-bit word stores the rank of $a_0 ... a_j$.
6. store precomputed ranks of *lc* in $\lceil \frac{X}{32} \rceil$ 8-bit words on positions a_{22}, a_{23}.
7. store precomputed ranks of *rc* in $\lfloor \frac{256-X}{32} \rfloor$ 8-bit words on positions a_{22}, a_{23} in a right-to-left order.

To calculate the number of occurrences in both proposed variants we do not need any additional popcount operation on bitvector data because of efficiently storing precomputed ranks of each 4-byte word of the bitvector in eight bits. This is because it is enough to store ranks just for the bitvector data in that chunk, whereas we store the count of occurrences for each character separately. Also now we can answer rank queries on *root*, *lc* and *rc* in that chunk with one cache miss at most.

The first proposed variant *wt_dna_il* has the chunk size equal to 768 bits. This is not a multiple of the cache line size, which is often 512 or 1024 bits. Thus $1024 - 768 = 256$ bits is prefetched in addition and will not be used in current calculation. The second variant *wt_dna_il_sa* has the chunk size equal to 1024 bits and the last 256 bits store the suffix array samples which can be potentially used in operation $locate(P)$. This also removes the possibility of another cache miss when accessing the sampled suffix array.

2.2 Precomputed k-mer Table

The precomputed k-mer table used recently in [13] skips first k operations of the searching phase in the FM-index therefore decreasing running time by storing precomputed values. We propose to store the values of k-mers in a lexicographical order. Then it is enough to store only the first value of range in a k-mer table, whereas the end range for $k\text{-}mer_i$ is the same as the start range for $k\text{-}mer_{i+1}$. The memory complexity is then for the DNA alphabet $O(4^k + 1)$, whereas we need to store end range for the last k-mer. Each k-mer value is stored on a position which equals to a following number notation of the DNA base: A = 00, C = 01, G = 10 and T = 11. Then for example 4-mer AAAA is stored on the position 0b00000000, TTTT on the position 0b11111111. In the last experiment we explore the use of the precomputed k-mer table for operation $count(P)$ with different values of k.

3 Experiments

Experiments were performed on personal laptop Asus F550V with 16 GB RAM DDR3 1600 MHz and CPU i7-7700HQ Kaby Lake with four 2,80 GHz cores. The OS was Linux (Ubuntu 18.04, 64 bit). Cache sizes were: L1 with 64 kB, L2 with 256 kB, L3 with 6 MB of memory. The size of the cache line was 64 bytes. The machine had no other significant CPU tasks running and only a single thread of execution was used during the experiments.

The program was compiled with Clang[2] compiler of version *6.0.0-1ubuntu2* and all source codes were written in the C programming language. The same compiler was used when compiling test programs implementing the calls of functions from SDSL. The compiler options in both cases were *-O3 -funroll-loops -fomit-frame-pointer -ffast-math*. In all tested variants of the FM-index the sampling size of the suffix array was 32.

We compared our proposed variants with other variants from SDSL:

- *v5Rank_to* - Huffman-shaped wavelet tree stored as bitvector with rank support $v5$, select support and text-order sampling of suffix array.
- *RRR15* - Huffman-shaped wavelet tree stored as RRR vector of size 15 with text order suffix array sampling strategy.
- *v5Rank* as *v5Rank_to* but with suffix array order sampling (the same method used in our variants).
- *vRank* as *v5Rank_to* but with rank support denoted in literature as v.
- *vRank_mcl* as *vRank* but with select support MCL.
- *il* - Huffman-shaped wavelet tree stored as interleaved bitvector with rank and select support.

Experiments consisted of running operations $count(P)$ and $locate(P)$. We measured the elapsed time of running these operations and also memory requirements of each version of the FM-index for the reference sequence. Times are

[2] https://clang.llvm.org/.

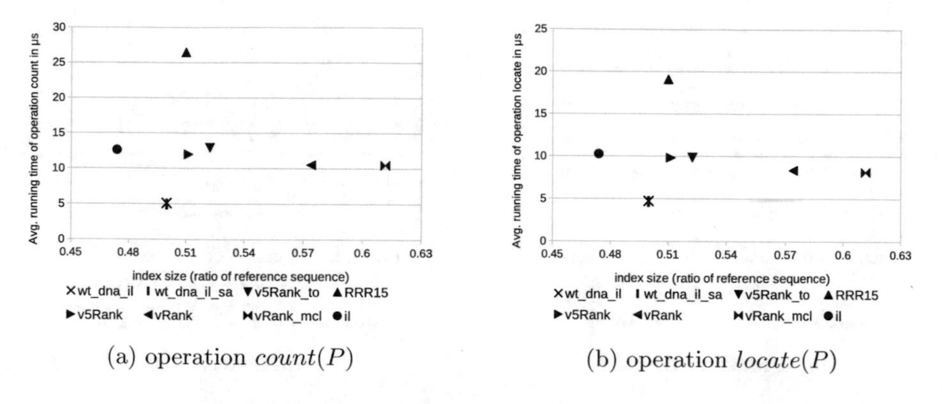

(a) operation *count*(P) (b) operation *locate*(P)

Fig. 1. Results on the *dna*200 dataset described precisely in text.

reported as the average elapsed time of running the operation count(P) per pattern and the operation locate(P) per an occurrence of the pattern. We used 2 different datasets.

In the first data set denoted as *dna*200, we used 200 MB version of *dna* testing data set from the Pizza & Chilli[3] as the reference sequence T, but we removed all characters but A, C, G and T. The total number of characters in the reference sequence was then 209,711,265. Ten millions of input patterns P of length 50 were taken from the random positions in the reference sequence to ensure that at least one occurrence of that input pattern exists in the reference sequence. There was a total of 22,693,775 occurrences of patterns.

As can be seen on Fig. 1 our variants achieved the shortest average time of running operations *count*(P) and *locate*(P) compared to other variants. There was no significant difference between running times of both our proposed variants. Our variants have also low memory requirements - around 50% of the reference sequence. Surprisingly running time of *wt_dna_il_sa* variant was not much shorter than *wt_dna_il* when locating a occurrence. This is probably due to the fact that the suffix array does not play big role when calculating the operation *locate*(P). Maybe the difference would be more noticeable if sampling size was smaller, however this would result in much higher memory requirements.

We experimented with a real-life dataset denoted as *bimb*, where as an input reference sequence we used the genome of *bombusimpatiens*, downloaded from NCBI[4] and again we removed all characters but A, C, G and T. The genome consisted then of 246,675,029 nucleotides. The input patterns P were of different lengths in this dataset. We experimented with pattern lengths of 30, 40, 50, 60, 70, 80, 90, 100 and 120. Also, in contrast with the previous dataset, the reads were generated randomly. A total of 2.75 millions of input patterns were generated for each length. There were a total of 64,196,519 occurrences in the text.

[3] http://pizzachili.dcc.uchile.cl/.
[4] www.ncbi.nlm.nih.gov.

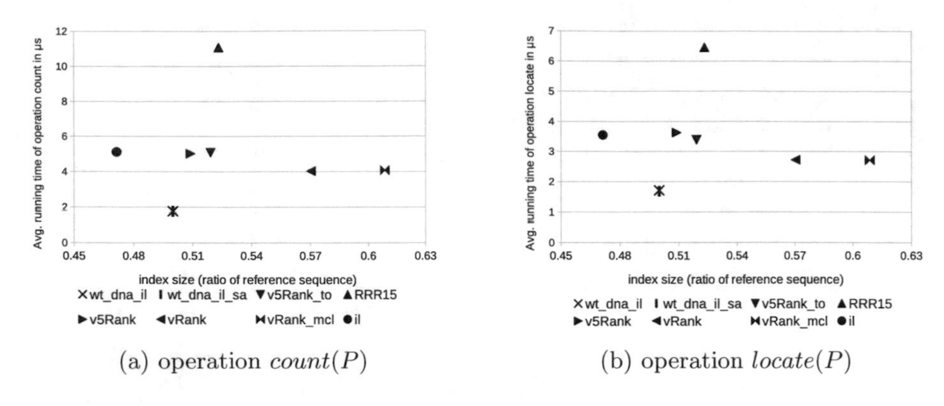

(a) operation $count(P)$ (b) operation $locate(P)$

Fig. 2. Results on the *bimb* dataset described precisely in text.

Figure 2 shows that compared to the first dataset there was no significant difference in running times nor memory requirements. *RRR15* had higher memory requirements than *v5Rank* and *v5Rank_to* compared to the previous experiment. Another interesting finding is that *v5Rank_to* achieved shorter running time of operation $locate(P)$ than *v5Rank* on this dataset.

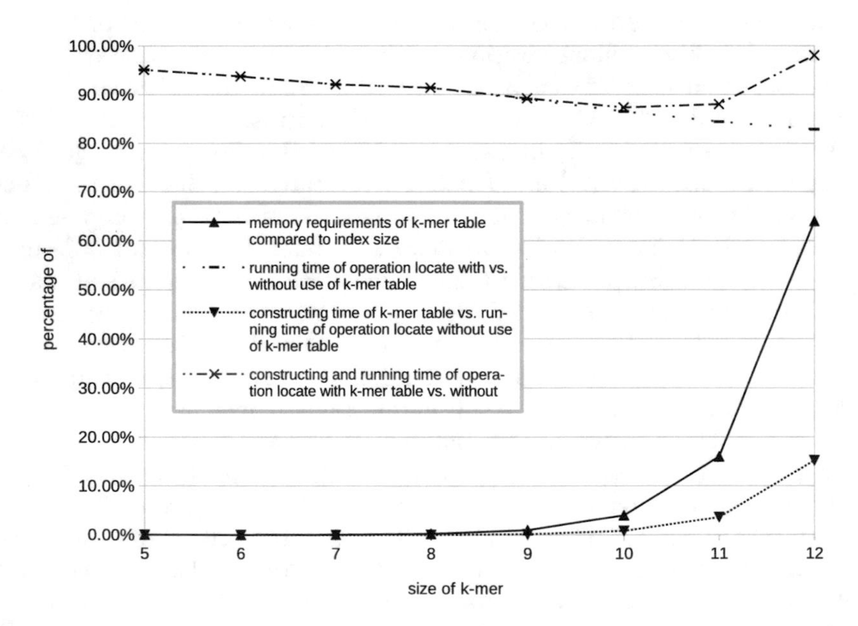

Fig. 3. Results of experiments using different values of k for the precomputed k-mer table.

In the last experiment we used the *dna*200 dataset to find an optimal value of k for the precomputed k-mer table. We used our variant *wt_dna_il_sa* with added k-mer table of different values of k. We measured the running time of operation $count(P)$, building time of the k-mer table and the memory allocated for the k-mer table. Figure 3 shows that the running time of operation $count(P)$ steadily decreases as k grows, however when $k = 11$, memory requirements and building time of k-mer table stopped being negligible. For $k = 12$ allocated memory for k-mer table were 64% of needed memory for index, whereas sum of building time of k-mer table and running time was almost the same as in the case when no k-mer table was used. Although memory requirements of the k-mer table depends only on the size of k, we could argue that with a longer reference sequence the needed memory for the k-mer table would be more negligible. The building time of the k-mer table did not grow as fast as the allocated memory with k, however this depends on the number of input patterns.

4 Conclusion

We presented cache-efficient FM-index variants to be used for reference sequences of the DNA alphabet. The proposed novel idea is based on interleaving all of the auxiliary data structures (sampled suffix array, wavelet tree bitvectors of BWT and rank support structures) of the FM-index in chunks of a fixed size. We implemented two variants, one interleaving all of the auxiliary data structures excluding the sampled suffix array, and the other one including also the sampled suffix array. Proposed variants were tested on large sample of patterns of different lengths. We compared our solution with other versions of the FM-index and the results have shown that our variants were at least two times faster when searching and locating patterns while having lower memory requirements than most of the other versions.

We have also experimented with a precomputed k-mer table of different values of k. The results have shown that the k-mer table can bring significant searching time reduction, whereas building time and memory requirements are negligible up to some value of k.

Acknowledgement. This work was partially supported by the Scientific Grant Agency of The Slovak Republic, Grant No. VG 1/0458/18, APVV-16-0484 and STU Grant scheme for Support of Young Researchers.

References

1. Adjeroh, D., et al.: The Burrows-Wheeler Transform: Data Compression, Suffix Arrays, and Pattern Matching. Springer, Boston (2008)
2. Altschul, S.F., Gish, W., Miller, W., Myers, E.W., Lipman, D.J.: Basic local alignment search tool. J. Mol. Biol. **215**(3), 403–410 (1990). http://www.sciencedirect.com/science/article/pii/S0022283605803602
3. Belazzougui, D., Navarro, G.: Alphabet-independent compressed text indexing. ACM Trans. Algorithms **10**(4), 23:1–23:19 (2014)
4. Berger, B., Peng, J., Singh, M.: Computational solutions for omics data. Nat. Rev. Genet. **14**(5), 333–346 (2013)

5. Burrows, M., Wheeler, D.: A Block-sorting Lossless Data Compression Algorithm. No. 124, Digital, Systems Research Center (1994)
6. Ferragina, P., Manzini, G.: Opportunistic data structures with applications. In: Proceedings 41st Annual Symposium on Foundations of Computer Science, pp. 390–398, November 2000
7. Ferragina, P., Giancarlo, R., Manzini, G.: The myriad virtues of Wavelet Trees. Inf. Comput. **207**(8), 849–866 (2009)
8. Ferragina, P., González, R., Navarro, G., Venturini, R.: Compressed text indexes: from theory to practice! CoRR abs/0712.3360 (2007)
9. Ferragina, P., Manzini, G.: Indexing compressed text. J. ACM **52**(4), 552–581 (2005). http://doi.acm.org/10.1145/1082036.1082039
10. Gog, S., Beller, T., Moffat, A., Petri, M.: From theory to practice: plug and play with succinct data structures. In: 13th International Symposium on Experimental Algorithms (SEA 2014), pp. 326–337 (2014)
11. Gog, S., Petri, M.: Optimized succinct data structures for massive data. Softw. Pract. Exper. **44**(11), 1287–1314 (2014)
12. González, R., Grabowski, S., Mäkinen, V., Navarro, G.: Practical implementation of rank and select queries. In: Poster Proceedings Volume of 4th Workshop on Efficient and Experimental Algorithms (WEA 2005), Greece, pp. 27–38 (2005)
13. Grabowski, S., et al.: FM-index for dummies. In: Kozielski, S., Mrozek, D., Kasprowski, P., Małysiak-Mrozek, B., Kostrzewa, D. (eds.) Beyond Databases, Architectures and Structures. Towards Efficient Solutions for Data Analysis and Knowledge Representation, pp. 189–201. Springer, Cham (2017)
14. Grossi, R., Gupta, A., Vitter, J.S.: High-order entropy-compressed text indexes. In: Proceedings of the Fourteenth Annual ACM-SIAM Symposium on Discrete Algorithms, SODA 2003, pp. 841–850. Society for Industrial and Applied Mathematics, Philadelphia (2003)
15. Grossi, R., Vitter, J., Xu, B.: Wavelet trees: from theory to practice. In: Proceedings of 1st International Conference on Data Compression, Communication, and Processing, CCP 2011, pp. 210–221, July 2011
16. Kärkkäinen, J., Puglisi, S.J.: Fixed block compression boosting in FM-indexes. CoRR (2011). http://arxiv.org/abs/1104.3810
17. Mäkinen, V., Navarro, G.: Run-length FM-index. In: Proceedings of DIMACS Workshop: "The Burrows-Wheeler Transform: Ten Years Later", pp. 17–19 (2004)
18. Mäkinen, V., Navarro, G.: Succinct suffix arrays based on run-length encoding. In: Proceedings of the 16th Annual Conference on Combinatorial Pattern Matching, CPM 2005, pp. 45–56. Springer, Heidelberg (2005)
19. Manber, U., Myers, G.: Suffix arrays: a new method for on-line string searches. In: Proceedings of the First Annual ACM-SIAM Symposium on Discrete Algorithms, pp. 319–327. Society for Industrial and Applied Mathematics, Philadelphia (1990)
20. Navarro, G.: Wavelet trees for all. J. Discrete Algorithms **25**, 2–20 (2014). 23rd Annual Symposium on Combinatorial Pattern Matching
21. Raman, R., Raman, V., Satti, S.R.: Succinct indexable dictionaries with applications to encoding k-ary trees, prefix sums and multisets. ACM Trans. Algorithms **3**(4) (2007). http://doi.acm.org/10.1145/1290672.1290680
22. Vigna, S.: Broadword implementation of rank/select queries. In: Proceedings of the 7th International Conference on Experimental Algorithms, WEA 2008, pp. 154–168. Springer, Heidelberg (2008)

A Parallel Implementation for Cellular Potts Model with Software Transactional Memory

A. J. Tomeu[1], A. Gámez[2], and A. G. Salguero[1(✉)]

[1] University of Cádiz, 11519 Puerto Real, CA, Spain
{antonio.tomeu,alberto.salguero}@uca.es
[2] IES Mar Mediterráneo, 04720 Aguadulce, AL, Spain
alexgamez@iesmarmediterraneo.org

Abstract. Cellular Potts Model is a mathematical model used to simulate biological systems in a wide scale range, from cells to organs. The model uses a Monte-Carlo approach to determinate for each cell, new state and actions like mitosis, movements or emission of pseudopods. Literature shows multiple implementations of CPM model, even incorporating parallel processing. These works use a data division approach that requires to take locks on data structures, or to spread information between tasks, slowing down simulations. This work proposes a fast implementation for CPM using software transactional memory to synchronize parallel tasks and to apply it to breast cancer in situ (DCIS). Execution times and *speedups* are calculated. Results show appreciable speedups.

Keywords: Cellular automaton · Cellular Potts Model · (Breast cancer in situ) DCIS · Gland · Locks · Multicore · Parallel programming · Shared memory · Software transactional memory · *speedup*

1 Introduction

The human breast is a gland with a high architectural complexity [13], where the parenchyma branches into pipeline networks of luminal cells formed in cross-section by an endoepithelium, surrounded by myoepithelial cells and basal membrane subsumed in the glandular stroma. Cellular alterations become a ductal carcinoma in *situ* (DCIS) in a first phase, where the neoplastic cells invade the light of the breast duct and become in an infiltrating ductal carcinoma in a second phase. At present it remains unclear how a DCIS is transformed into an infiltrating carcinoma, even though the presence of mutations in BRCA1, BRCA2, PTEN and PT53 is one of the known causes. A better knowledge of the DCIS phase is necessary and the mathematical modelling is not a newcomer in this stage [1, 2, 5, 8, 10] since it allows to perform a multitude of *in silico* experiments with almost negligible cost, leading to the laboratory (in *vitro* and in *vivo* experimentation) only the most promising ones. One of the mathematical models used for the DCIS computational research is the Cellular Potts Model (hereafter referred to as CPM) [3, 7, 16] which uses a reticular approach to model a histological section of a human breast duct. This work presents a parallel implementation of the CPM as alternative to those already known in literature [3], that

F. Fdez-Riverola et al. (Eds.): PACBB 2019, AISC 1005, pp. 53–60, 2020.
https://doi.org/10.1007/978-3-030-23873-5_7

differs from them by the use of control techniques based on the transactional memory software [17], to manage the data grid parallel management that simulates the neoplasm. Thus, we implemented the proposed parallel model using the Java programming language and then we made time and *speedup* measurements of several nodes in the cluster of processors of our University. Finally, we discussed the results and we proposed our conclusions.

2 Biology of Adenocarcinomas

The adenocarcinomas in general, and the breast cancer in particular, are gestated inside of the tubules of the gland. In the case of the human breast and in a normal state, the histology [13] shows that these ducts are composed of two layers of cells: The innermost layer formed by luminal cells which is evolved by a second layer of myoepithelial cells, wrapped by a basement membrane [6, 13]. Both types of cells are derived from a single class of stem progenitor through a differentiation mechanism that starting from stem cell, it allows the existence of two germ lines that conclude in differentiated cells either myoepithelial or luminal.

Breast ductal adenocarcinomas begin (Fig. 1) as one (or several mutations) in the genomes of these cells. Although ultimately the luminal cells are where the luminal neoplastic transformation takes place, which initially invade the light of the duct breaking the normal double-layer architecture (hyperplasia) then go by filling out the light of the duct (ductal carcinoma in situ (DCIS) and finally breaking the basement membrane and invading the glandular parenchyma; At this point the disease acquires an infiltrative characteristic, appears to be the possibility of metastatic processes by hematogenous routes and/or lymphatic, and morbidity and mortality among the sick and treatment costs rise considerably.

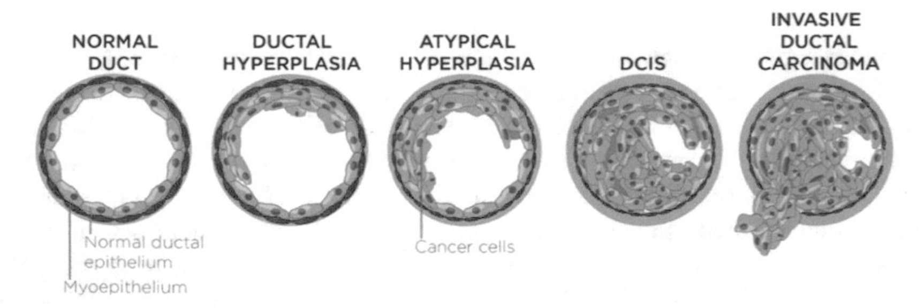

Fig. 1. Natural history of breast adenocarcinoma.

The purpose of the CPM model of the disease is to simulate the evolution of the same from a structurally normal duct, until a passage where some cells have undergone malignant transformations in their genome, have begun to proliferate without control, and have filled the light of the conduit.

3 Cellular Potts Model (CPM)

Also known as Glazier-Graner-Hogeweg model, CPM provides a framework [1, 2, 7] allowing complex simulations of cellular behaviors, at different levels of structural complexity [7, 10], ranging from the cell organelles to whole organs. It also allows to accommodate a wide range of cell processes [7]. In this model, a cell behaves as a function of a balance of forces described by energy *H*, which is the sum of the interface energies with other cells and the deviation of the volume of that cell from its baseline value (e.g. when a pseudopod issuing). In the case of DCIS [14], the CPM allows the simulation of tumor growth in situ by defining a 2D grid ζ with null borderline condition [3–5, 12, 15, 17]. Each node of the grid is defined by a pair of coordinates (i,j), and a symbol *k* taken at some finite alphabet of symbols $\sum$. A cell can be defined as a subset S = {$\alpha = (i, j, k)$: *k* is constant} of ζ which, depending on the scale of the model, can represent a cellular organelle, a cell or a tissue. The model evolves in discrete-time stages. In each of them, a node of the grid tries to alter its location. A stochastic Monte-Carlo process is used for [3–5, 7, 8, 12, 15, 16] so that each reticle node can try to change its location interacting with the neighboring cells. This attempt meets according to the transition function described below (Eq. 1):

$$P(\alpha(i,j,k) \rightarrow \alpha(i',j',k')) = \begin{cases} e^{-\frac{\Delta H}{T_m}} & if\ \Delta H > 0 \\ 1 & if\ \Delta H \leq 0 \end{cases} \tag{1}$$

Where, ΔH is the change in the effective energy, according to the Hamiltonian *H* as defined below (Eq. 2), T_m is the temperature parameter and the triplet (i, j, k) specifies a node of the grid that is part of a concrete cell domain α. The Hamiltonian *H* is defined by the following equation:

$$\begin{aligned} H = & \sum_{(i,j,k),(i',j',k')} J_{\tau(\alpha(i,j,k)),\tau(\alpha(i',j',k'))}\left(1 - \delta_{\alpha(i,j,k),\alpha(i',j',k')}\right) \\ & + \sum_{\alpha} \lambda_V(\tau)(V(\alpha) - V_t(\alpha))^2 + \sum_{\alpha} \lambda_S(\tau)(S(\alpha) - S_t(\alpha))^2 \end{aligned} \tag{2}$$

Here, τ represents the type of agent. In this case, when applied to the DCIS, agents will represent 3 different cell types: luminal cells, extracellular matrix cells and myoepithelial cells, modeled by an alphabet $\sum$ projected on $\mathbb{Z}_3$. In the equation defining *H*, the first term describes the energy of adhesion between a cell and its neighbors. The second term defines the volume and the degree of compressibility of the cell. For this, the difference between the volume (V) and the target volume (V_t) is multiplied by λ_v parameter that describes the stiffness of the agent. The third Hamiltonian term comes to model the cell elasticity of the cell (S) and (S_t) represents the surface and the target surface as a function of it. The model begins with an initial structure similar to a cross-sectional layer of a normal duct. For this purpose [6, 13] approximately 50 luminal cells (L) are surrounded by an outer layer of myoepithelial cells (M); both layers are built with a circular geometry. In the simulation of this initial state, each cell begins as a domain square of 10 $\times$ 10 nodes with a *k* common value to all them immersed in the ζ grid. The λ_V volumetric parameters allow to define the elasticity of the cellular cytosol. Since in this work the focus is mainly on the acceleration of the computations of the CPM, we have chosen their values based on previously published works [3] that place more emphasis on the model fidelity then the modeled biological reality.

4 The Parallel Implementation of the Model

In order to accelerate the CPM simulation, we have performed an implementation that processes with parallel tasks the tissue domain ζ in which, unlike other proposals [3, 5, 7, 9, 11, 12, 14, 15, 17], the data structure is not divided between tasks, since this forces to propagate information among threads that are responsible to process bordering substructures when a cell crosses the border between nodes as a consequence of a mitosis or of the extension of a pseudopod.

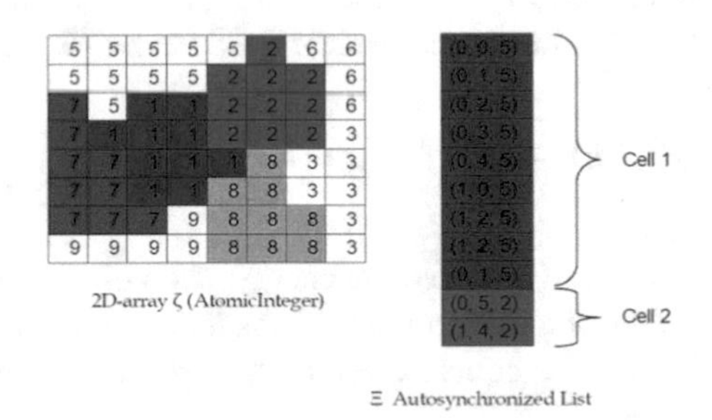

Fig. 2. Data Structures used in our implementation of the CPM. The 2D array models the simulated tissue domain ζ; the list Ξ stores the cells that have already tried their process of expansion, and managed under transactional memory.

Instead, we develop the stochastic modeling phase by dividing the number of iterations between stochastic subprocesses (CPM algorithm as reference) independent of each other where each of which comes to use different random generators, with different initial seeds generated from the system clock. Then the implementation proposed in this work uses the following data structures (shown in Fig. 2): a 2D-array is used to represent the tissue domain ζ. In the ζ grid, each node is linked to a k numerical value that indicates the cell that occupies it. A collection of related S nodes, with the same numerical value associated to define a cell; a list Ξ of (i, j, k) elements which contains the information of those nodes of the ζ grid that already tried the expansion process described by Eq. (1). Each element of the list defines a node (i, j) of the ζ grid along with the k numerical value that indicates which cell occupies it. If there is none, then $k = 0$. Parallel tasks are coded using the Java Runnable interface, and were processed through a *pool* of threads, delegating on it the entire life cycle management of the tasks. Each parallel task runs the Evolve algorithm which is illustrated below. All of them were given access to the common data structures ζ and Ξ through specific references. The "Evolve" is performed by all parallel tasks, is a direct implementation of CPM basic algorithm, which stochastically selects (lines 1.1 and 1.2) a node of the grid (which is part of a cell) that has not the opportunity to change its location.

```
Algorithm Evolve(ζ, Ξ)
1.for(i=1, i<niterations/ntasks, i++){
        1.1 x=random(xmax);
        1.2 y=random(ymax):
   2. runInTransaction(){if !((x,y) in Ξ){
          cell= ζ[x][y];
          Ξ.add((x,y,k));}
      else goto 1.1
     }
   3. neighbourX = getRandom(rangXMin,rangXMax);
   4. neighbourY = getRandom(rangYMin,rangYMax);
   5. cellNeighbour = ζ [neighbourX][neighbourY];
   6. J=getEnergyAdhesionForNeighbors(cell);
   7. V=lambdaV*getVolumeForCell(cell);
   8. S=lambdaS*getSurfaceCell(cell);
   9. deltaH=J+V+S.
  10. if(deltaH>0)
          P=Math.exp(-eltaH/params.getTemperature());
     else if(deltaH<=0)P=1;
  11. if(p>=random()) ζ[x][y]=cellNeighbour();
}
```

Once a node is chosen, it spreads such information to all other threads, by inserting it into the Ξ list within a transaction. Since the Ξ list is processed within a transaction [17] in line 2, any race condition between threads that attempt to relocate to the same cell node is solved in the verification of the bifurcation of the line 2, which is processed atomically. The thread that manages to complete its transaction (*commit*) updates the list and the rest undo the transaction (*roll-back*) and start it again with an updated view of the Ξ list. Since there is now a single shared data structure (the Ξ list, Fig. 5) instead of the $n - 1$ contact areas than usual division domain data between n tasks requires [7, 9, 15, 17] the containment of the parallel threads is reduced. From here, the different terms of the Hamiltonian H are determined (lines 6, 7 and 8), the probability of location change or mitosis (conditional bifurcation of line 10) is computed and, depending on it, this change occurs or not (line 11). For proper parallel tasks operation model it is necessary to impose certain security conditions that limit its computational efficiency but that allow to guarantee an adequate use of the data structures between the parallel threads that process them. The 2D-array ζ which has been implemented using a standard array of bytes in Java, must be secured against writing over specific positions of the array. It is achieved by forcing the threads to previously pass through the Ξ list, which does not allow two threads to write at the same point of ζ and is protected within a transaction. We have chosen this option instead of the usually approach proposed in the literature (locks on the array or propagation of information) because it is proven that this approach presents better response times than the domain data division among the threads [3, 5, 9, 12, 15, 17] and the control of the borders through locks, which also requires an algorithm to disseminate the information to bordering data areas. The Ξ list needs to be accessed within a transaction [17], so when a thread checks if the node that intends to develop (line 2 of the algorithm), it does with the information updated (Fig. 3). Then among the multiple options that the Java API offers to implement Ξ we have chosen the `ArrayList` class processed within a transaction which guarantees that all read/write operations are performed on an updated copy of the underlying array, thus being secure against parallel threads.

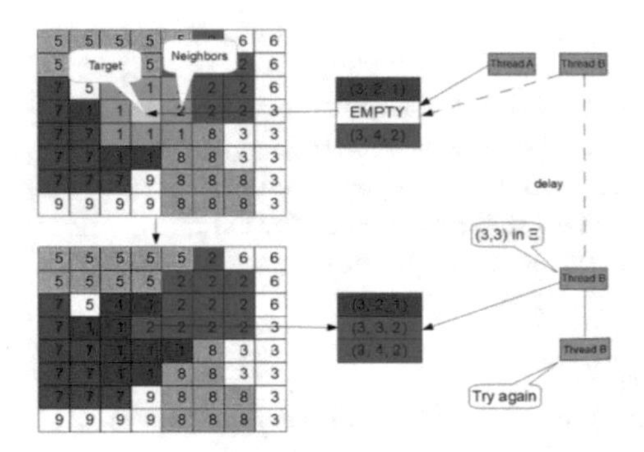

Fig. 3. Parallel threads in simultaneous access to the list and its resolution through a transaction. Thread A validates its transaction through *commit*, whereas Thread B must undo its transaction through *roll-back*, and try again, to finish updating a different cell.

5 Measures and Discussion of Results

In our experiments, the transactional memory software was supported by the Clojure Programing Language extension for Java [17]. Each node of the cluster of our university has two Intel® Xeon™ E5 (2.6 GHZ) processors, that yield 20.8 GFLOPS in total, with 128 GB of RAM and without *hyperthreading*.

The entry node operates with *HP Cluster Management Utility* on *Red Hat Enterprise Linux* for HPC. The Java development kit version was *Oracle 1.8.0.151-1.b12. el7_4*. The times measures were taken on 8 dedicated nodes of the cluster, for one task (sequential algorithm) and for a number of tasks increasing up to 12. In all cases the size of the ζ grid had 900 × 900 cells. The results (average times ± SD) are illustrated in Fig. 4. It is observed how the proposed parallel implementation succeed to reduce the run time to a minimum average value of 2.285 s for 12 tasks. For comparison purposes, if the list Ξ is managed with locks-based techniques, a minimum time (for 12 tasks) of 6.32 s is obtained. Subsequently, the resulting *speedups* were calculated whose mean values (±SD) are shown in Fig. 5. It is observed how the *speedup* behaves inversely to time, growing as the number of threads increases and again reaching its optimal value for 12 tasks with an average value of 10.76 over a theoretical maximum of 16, which is more than reasonable. All average values are taken on data from 8 different nodes of our cluster.

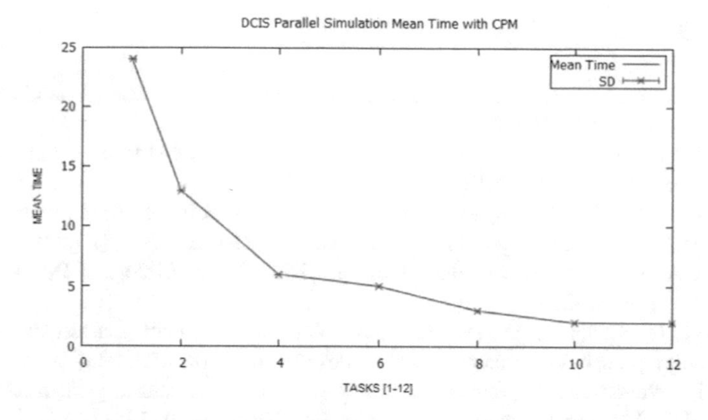

Fig. 4. The average run times along with their standard deviation.

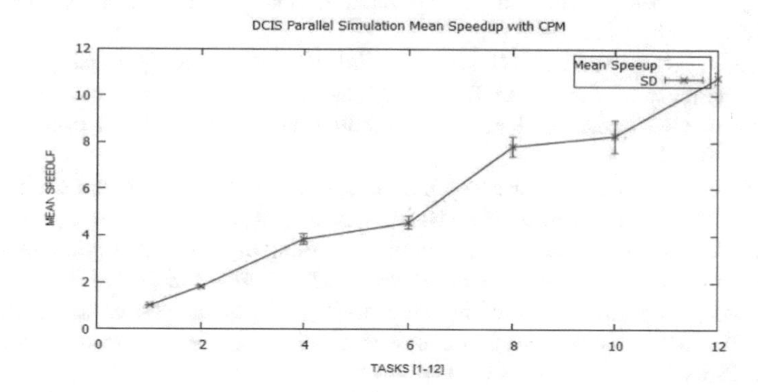

Fig. 5. Speedups average along with its standard deviation.

6 Conclusions and Future Work

In this work, we present a fast CPM implementation that adds to the ζ grid that models the domain ζ an auto-synchronized list Ξ, based on transactional memory that contains the cells in the grid that have already been processed, and parallelizes the basic CPM routine among multiple parallel threads. The Ξ list acts as a synchronizing pattern among threads, and as a filter to access to the ζ grid, only allowing it when it comes to compute a cell that has not yet been processed, and always within a transaction. Tests show competitive execution times. Our future work will be oriented toward the application of our fast implementation of the CPM proposal to glandular tumors different to the DCIS, as those of prostate or thyroid, and to the implementation of the model on a GPU architecture. At the same time we are also interested in exploring the implementation of the model using approaches with actors and distributed agents.

References

1. Altrock, P., Liu, L., Michor, F.: The mathematics of cancer: integrating quantitative models. Nat. Rev. Cancer **15**, 730–745 (2015)
2. Byrne, H.: Dissecting cancer through mathematics: from the cell to the animal model. Nat. Rev. Cancer **10**, 221–230 (2010)
3. Chen, N., et al.: A parallel implementation of the Cellular Potts Model for simulation of cell-based morphogenesis. Comput. Phys. Commun. **176**(11–12), 670–681 (2007)
4. Deutsch, A., Dormann, S.: Cellular Automaton Modeling of Biological Pattern Formation. Birkhäuser, Boston (2005)
5. Enderling, H., Chaplain, M., Anderson, A., Vaidya, J.: A mathematical model of breast cancer development, local treatment and recurrence. J. Theor. Biol. **246**(2), 245–259 (2007)
6. Erbas, B., Provenzano, E., Armes, J., Gertig, D.: The natural history of ductal carcinoma in situ of the breast: a review. Breast Cancer Res. Treat. **97**(2), 135–144 (2006)
7. Graner, F., Glazier, J.A.: Simulation of biological cell sorting using a two-dimensional extended Potts model. Phys. Rev. Lett. **69**(13), 2013–2016 (1992)
8. Gevertz, J., Gillies, G., Torquato, S.: Simulating tumor growth in confined heterogeneous environments. Phys. Biol. **5**(3) (2008)
9. Giordano, A., et al.: Parallel execution of cellular automata through space partitioning: the landslide simulation Sciddicas3-Hex case study. In: Proceedings of 25th Euromicro International Conference on Parallel, Distributed and Network-Based Processing (PDP), pp. 505–510 (2017)
10. Kang, S., et al.: Biocellion: accelerating computer simulation of multicellular biological systems models. Bioinformatics **30**(21), 3101–3108 (2014)
11. Kam, Y., Rejniak, K., Anderson, A.: Cellular modeling of cancer invasion: integration of in silico and in vitro approaches. J. Cell. Physiol. **227**(2), 431–438 (2012)
12. Monteagudo, A., Santos, J.: Studying the capability of different cancer hallmarks to initiate tumor growth using a cellular automaton simulation. Application in a cancer stem cell context. Biosystems **115**, 46–58 (2014)
13. Kumar, P.: Manual of Practical Pathology. CBS Publisher & Distributors P Ltd. (2011)
14. Norton, K., Wininger, M., Bhanot, G., Ganesan, S., Barnardh, N., Shinbrotb, T.: A 2D mechanistic model of breast ductal carcinoma in situ (DCIS) morphology and progression. J. Theor. Biol. **263**(4), 393–406 (2009)
15. Salguero, A., Capel, M., Tomeu, A.: Parallel cellular automaton tumor growth model. Advances in Intelligent Systems and Computing (AISC), vol. 803 (2018). https://doi.org/10.1007/978-3-319-98702-6_21
16. Scianna, M., Preziosi, L.: Cellular Potts Models: Multiscale Extensions and Biological Applications. Mathematical and Computational Biology Series. Chapman & Hall/CRC, Boca Raton (2013)
17. Tomeu, A.J., Salguero, A.G., Capel, M.: Speeding up tumor growth simulations using parallel programming and cellular automata. IEEE Lat. Am. Trans. **14**(11), 4603–4619 (2016)

Inferring Positive Selection in Large Viral Datasets

Hugo López-Fernández[1,2,3,4,5], Pedro Duque[1,2], Noé Vázquez[3], Florentino Fdez-Riverola[3,4,5], Miguel Reboiro-Jato[3,4,5], Cristina P. Vieira[1,2], and Jorge Vieira[1,2](✉)

[1] Instituto de Investigação e Inovação em Saúde (I3S), Universidade do Porto, Rua Alfredo Allen, 208, 4200-135 Porto, Portugal
pedro.duque@i3s.up.pt

[2] Instituto de Biologia Molecular e Celular (IBMC), Rua Alfredo Allen, 208, 4200-135 Porto, Portugal
{cgvieira, jbvieira}@ibmc.up.pt

[3] ESEI, Department of Computer Science, University of Vigo, Campus As Lagoas, 32004 Ourense, Spain
nvazquezg@gmail.com,
{hlfernandez, riverola, mrjato}@uvigo.es

[4] The Biomedical Research Centre (CINBIO), Campus Universitario Lagoas-Marcosende, 36310 Vigo, Spain

[5] SING Research Group, Galicia Sur Health Research Institute (ISS Galicia Sur), SERGAS-UVIGO, Vigo, Spain

Abstract. A large amount of viral nucleotide sequences is available in databases that can be used to identify positively selected amino acid sites, and thus make inferences on which sites are important for immune system escape and adaptation to their host. Nevertheless, the software pipelines needed to analyse such large datasets usually imply long running times. Moreover, their power to identify positively selected amino acid sites may not be similar. Therefore, here we first analyse, under a variety of conditions, the performance of different software applications and then propose a protocol for the analysis of large datasets.

Keywords: Viruses · Positively selected amino acid sites · Big data

1 Introduction

Nowadays a large amount of viral sequences is available in large databases, such as the ViPR[1] (Virus Pathogen Resource). In this database, by January 2019, data was available for 17 viral Families, belonging to 148 genera, and 5773 species (622,827 strains), represented by 821,914 GenBank sequences. This data can be used to identify positively selected amino acids (those responsible for adaptive features such as host immune response escape), as previously done for a few viruses and coding sequences

[1] https://www.viprbrc.org.

F. Fdez-Riverola et al. (Eds.): PACBB 2019, AISC 1005, pp. 61–69, 2020.
https://doi.org/10.1007/978-3-030-23873-5_8

(see for instance, [1–3]). The positively selected amino acid sites are usually identified using maximum-likelihood methods based on models of codon substitution, such as codeML [4] or FUBAR [5]. It should be noted that, being phylogenetic methods, they assume that there are no recombinant sequences in the dataset. Shriner et al. [6] and Anisimova et al. [7] have shown that recombination (especially high levels) can lead to the detection of many false positively selected amino acid sites when using codeML. The more realistic M7 and M8 codeML models are less sensitive to recombination than models M1a and M2a [5, 7], but require longer execution times. FUBAR has been presented as a much faster alternative to codeML but the power of the two methods to detect positively selected amino acid sites has not been directly compared. One alternative to phylogenetic methods is OmegaMap, that uses a population genetics approximation to the coalescent with recombination to identify positively selected amino acid sites [8]. Nevertheless, it assumes that the set of sequences being analysed is a population sample, which in the case of viruses may be difficult to define.

The computer application ADOPS [9], allows running in an automated way all the steps needed to infer positively selected amino acid sites when using codeML [4], starting from a FASTA file with non-aligned coding sequences. Moreover, it is able to run in batch mode, meaning that, it is practical to run large scale projects. A Docker image for ADOPS is available at the Bioinformatics Docker Images Project[2], meaning that it is now possible to run ADOPS without having to install the multiple software packages that are required by this application. The resulting ADOPS projects, containing all information related to the execution of the projects, can be stored in the B+ database [10]. At this website, all project details can be viewed without the need to download them. Projects can also be downloaded, the data reused, and new analyses performed (including sequence removal and addition of novel sequences). These tools are extremely useful when conducting large-scale projects.

Here, we report the results of a case study, involving the Zika, Dengue, Ebolavirus, and HIV1 virus sequences, aimed at identifying the main features that a protocol should have in order to have a good power to detect positively selected amino acid sites and at the same time avoid computer time waste. We used four unrelated viruses to increase the chance of developing a protocol that is suitable for a wide variety of viruses. We discuss sequence selection, recombination detection, power to detect positively selected amino acid sites, and running times. Then, we propose a protocol for analysing a large number of viruses, such as those available at the ViPR (see Footnote 1) database.

2 Materials and Methods

2.1 Datasets

Coding sequences were downloaded from the ViPR (see Footnote 1) database, as well as from the HIV database[3], during 2017/2018, and processed using SEDA[4] [11], to

[2] https://pegi3s.github.io/dockerfiles/.

[3] https://www.hiv.lanl.gov.

[4] https://www.sing-group.org/seda/.

eliminate all sequences presenting ambiguous positions or in-frame stop codons (Supplementary Table 1; https://doi.org/10.5281/zenodo.2560649). Then, for each virus and coding sequence, SEDA was also used to obtain two datasets, one containing sequences that are different at the nucleotide level (referred from now on as nucleotide dataset), and another one containing sequences that are different at the amino acid level (referred from now on as amino acid dataset). This way we can test whether using sequences that are different at the nucleotide or amino acid level leads to different results. Since different coding regions present different sequence numbers and variability levels, the number of remaining sequences is different for each coding region.

It is also unclear whether increasing sequence number always leads to an increased power to detect positively selected amino acid sites. Therefore, for Zika and Ebolavirus the maximum number of sequences used was 100, while for Dengue and HIV1 was 50. Both numbers are commonly seen in studies aimed at the identification of positively selected amino acid sites. In every case, five independent replicates were used to determine whether a single sample of a given size is able to capture all positively selected amino acid sites. The generation of random samples of a given size was also performed using SEDA [11].

2.2 Recombination and Positive Selection Analyses

When using phylogenetic methods, recombination can lead to the detection of false positively selected amino acid sites [6, 7], and thus we searched for evidence of recombination using the sets of unique sequences and the Phi test [12], as implemented in SplitsTree [13]. If no significant evidence for recombination is found, then we run codeML [4], as implemented in ADOPS [9], otherwise we run OmegaMap [8]. Nevertheless, for the Zika amino acid datasets showing evidence for recombination, we run both methods to have insight into how codeML behaves in the presence of some recombination. Sequences were aligned at the amino acid level using Muscle [14], and the corresponding nucleotide alignment obtained. Then, phylogenies were inferred using MrBayes [15], using only codons that are aligned with a confidence score of three or higher. The model of sequence evolution used was the GTR (allowing for among-site rate variation and a proportion of invariable sites). Third codon positions were allowed to have a gamma distribution shape parameter that is different from that for first and second codon positions. Two independent runs of 2,000,000 generations with four chains each (one cold and three heated chains) were used. Trees were sampled every 100th generation and the first 2500 samples were discarded (burn-in). The average standard deviation of split frequencies and the potential scale reduction factor for every parameter showed that convergence has been achieved. Positively selected amino acid sites are then inferred using codeML. All operations are performed using ADOPS, and all results stored at the public B+ database[5] [10] (project numbers BP2017000007, BP2017000008, BP2017000012 and BP2018000001). It should be noted that the ADOPS and B+ layout was changed to include two extra tabs: (1) a tab called OmegaMap Summary showing the raw results of the OmegaMap runs; (2) a tab

[5] http://bpositive.i3s.up.pt/.

named PhiPack Log showing the results of the analyses performed for each dataset using the PhiPack[6] software application, that implements the Phi test, as well as two others (NSS and Max Chi^2). If positively selected amino acid sites are detected using OmegaMap they will also be shown in the PSS tab. In order to compare results across replicas, a sequence from each sequence alignment was taken and aligned to a reference sequence.

OmegaMap [8] and FUBAR [5] were run using the default parameter values and the sequence alignments generated by ADOPS [9], as described above. FUBAR was run as implemented in the DataMonkey [16] platform. Statistical analyses were performed using the IBM SPSS[7] software package.

3 Results and Discussion

For the Ebola L amino acid and nucleotide datasets, codeML [4] produced an error, which was not due to physical memory or processor limitations. The value of the number of sequences times the number of ungapped codons (from now on referred as *p*) for these datasets is 187,566 and 217,900, respectively, much higher than that obtained for all other remaining datasets for which the highest value is 89,000. This suggests that there is a limit regarding the value of *p* that codeML can handle. A similar behaviour has been previously reported by Murrell et al. [5], when attempting to analyse a set of 476 reverse transcriptase HIV1 sequences (335 codons) with a *p* value of 159,460. We thus recommend using a maximum value of *p* of around 90,000.

As shown in Supplementary Fig. 1 (https://doi.org/10.5281/zenodo.2560649), running times seem to be linearly correlated with *p*. Nevertheless, both the slopes and the fit to the data depend heavily on the virus analysed. For instance, for a value of *p* of 50,000, codeML [4] model M8 running time predictions are 8.2, 27.8, 13.8, and 6.6 h for Zika, Dengue, Ebolavirus, and HIV1, respectively. Differences in variability levels among the datasets could, in principle help explain this observation. Surprisingly, overall variability levels (estimated by the PhiPack (see Footnote 6) software application) are negatively correlated, or not correlated at all, with running times (Supplementary Fig. 2; https://doi.org/10.5281/zenodo.2560649). This could be due to a negative correlation between protein coding region length and variability levels (Supplementary Fig. 3; https://doi.org/10.5281/zenodo.2560649). When performing stepwise regression analyses, using both *p* and variability level as independent variables, only *p* is included in the final models.

Although codeML [4] model M8 takes longer to run than model M2a, we performed both the M8–M7 and M2a–M1a comparisons under the assumption that model M8 is better suited to detect positively selected amino acid sites. This is indeed the case since there are 23 and 21 runs where M8 identifies more positively selected amino acid sites than model M2a when considering the nucleotide and amino acid datasets, respectively (Supplementary Table 2; https://doi.org/10.5281/zenodo.2560649). This

[6] https://www.maths.otago.ac.nz/~dbryant/software/PhiPack.tar.

[7] https://www.ibm.com/analytics/spss-statistics-software.

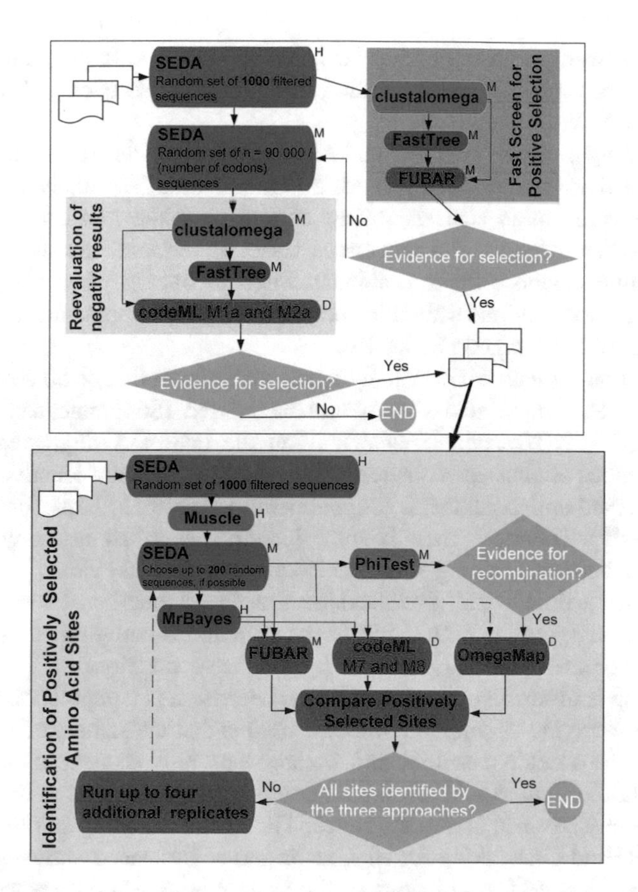

Fig. 1. Schematic workflow of the proposed protocol. D – days; H – hours; M – minutes.

agrees with previous reports that models M7 and M8 are more realistic [5, 7]. On average, model M8 detects 5.5% more positively selected amino acid sites than model M2a. In the case of two replicas for both the nucleotide (N-HIV1-VPR-4 and N-Ebola-NP-1) and amino acid (A-HIV1-VPR-4 and A-Zika-NS4B-1) datasets, using only M2a, would mean missing positively selected amino acid sites.

For five cases for both the nucleotide (NEF-2 and 3, and ENV-1, 3 and 4) and amino acid (VIF-5, TAT-5 and ENV-3, 4 and 5) HIV-1 datasets (Supplementary Table 2; https://doi.org/10.5281/zenodo.2560649), codeML [4] model M2a reports more positively selected amino acid sites than model M8. These additional positively selected amino acid sites identified with M2a could be false positives, since Murrell et al. [5] have shown that model M2a is sensitive to slight model misspecifications, such as assuming that all positively selected amino acid sites are subjected to similar selection strength. When using the amino acid datasets, on average, 5.6% more positively selected amino acid sites are detected than when using the nucleotide datasets but this difference is not statistically significant (Non-parametric Sign test; $P > 0.05$). It should be noted that, for the datasets where the same number of sequences was analysed when using both the amino acid and the nucleotide protocol (Dengue and HIV1), there are no differences regarding variability levels averaged across replicas (Sign test; P > 0.05).

Besides being more realistic, M7 and M8 codeML models are also less sensitive to recombination than models M1a and M2a [7], and thus less likely to identify false positively selected amino acid sites. This should be taken into account because, in many cases, when using the full sequence dataset no evidence for recombination was recovered, while evidence for recombination was recovered when using the random subsets of sequences and the same Phi (permutation) test (Supplementary Tables 3–6; https://doi.org/10.5281/zenodo.2560649).

It is clear that, at least for HIV1, by using a single replica of 50 sequences, many positively selected amino acid sites would be missed (Supplementary Tables 7–15; https://doi.org/10.5281/zenodo.2560649). Even the use of a single replica with the maximum number of allowed sequences given *p* (100), does not identify the full set of positively selected amino acid sites (Supplementary Table 15; https://doi.org/10.5281/zenodo.2560649). Therefore, there is some justification to use as many sequences as possible, even because, judging from the Zika virus datasets, where *p* is close to the limit, the running time is still predicted by the linear function. However, given the differences found regarding both the fit to the data and running times, it could be that the runtime is greater than that predicted by the linear function.

OmegaMap is insensitive to recombination because it is a population-based method [8]. The most worrying assumption of the method is that the sample of sequences to be analysed must be a random sample, and we are using only sequences that are different at the nucleotide or amino acid level. This could conceivably lead to the identification of false positively selected amino acid sites. Therefore, we compared the performance of codeML [4] and OmegaMap for all Zika amino acid datasets showing evidence for recombination. Despite the violations of the assumptions of both approaches, neither method detected positively selected amino acid sites in these datasets. OmegaMap running times are, however, at least one order of magnitude longer than those for codeML, and can be predicted with high confidence using a power function (Supplementary Fig. 4; https://doi.org/10.5281/zenodo.2560649).

The high mutation rate usually exhibited by viruses could in principle also lead to the identification of false positively selected amino acid sites, but in our analyses a similar number of such sites was identified in the Zika and Ebolavirus datasets, despite very different total variability levels (Supplementary Table 2 and Supplementary Fig. 2; https://doi.org/10.5281/zenodo.2560649).

For large datasets, FUBAR is orders of magnitude faster than codeML when using model M8 [5]. Moreover, it allows the analysis of datasets with thousands of sequences, while there seems to be a limit for codeML [4] for a *p* value of about 90,000. Nevertheless, they use different approaches for identifying sites under positive selection, and thus can produce different results. When different methods fail to identify the same positively selected amino acid sites, the confidence is necessarily lower than when there is an agreement. Therefore, when analysing large datasets what is needed is a way of quickly identifying genes that might show positively selected amino acid sites, and thus should be subjected to further analyses. Murrell et al. [5] report that for a *p* value of 159,460 (the HIV1 reverse transcriptase dataset) FUBAR only takes 278 s (partly because it can use several processors; the reported running time is when using three dedicated processors), and thus this could be the logical choice for making a fast screen for those genes that might show evidence for positively selected amino acid sites, using fast alignment methods such as clustalomega [17] and fast tree building methods, such as FastTree [18] that, although much faster than RAxML (the current leading method for large-scale maximum likelihood estimation), shows little or no degradation in tree accuracy, as compared to RAxML [19]. Nevertheless, it must be first shown that the power to detect positively selected amino acid sites is similar in FUBAR and codeML. Therefore, we compared the results obtained using FUBAR, when using the sequence alignment obtained for the HIV1 dataset N_ENV_1_1 (50 sequences). Of the 64 positively selected amino acid sites identified by codeML model M8, only 13 were identified by FUBAR. This could suggest that when using a relatively small number of sequences, FUBAR has less power to detect positively selected amino acid sites than codeML model M8. Nevertheless, FUBAR also identified seven additional amino acid sites that are identified as positively selected by model M8, when using other replicas for the same gene. Therefore, in some cases, FUBAR can use the information contained in the sequence alignments more efficiently. Finally, 21 sites were also identified by FUBAR that were not identified in any of the five codeML model M8 replicas. Whether these are false positively selected amino acid sites remains to be determined. Given this behaviour, we next proceeded to determine whether, for datasets with less than 250 sequences in total, FUBAR would detect positively selected amino acid sites in the datasets where codeML model M8 did so (Zika NS4B and NS5; Ebolavirus NP and sGP). FUBAR identified positively selected amino acid sites for Zika NS4B and NS5 but failed to detect positively selected amino acid sites for Ebolavirus NP and sGP (using either the nucleotide or amino acid dataset).

We also addressed whether FUBAR [5] would falsely identify positively selected amino acid sites in the presence of recombinant sequences. Therefore, we run FUBAR, using the full dataset of different sequences for those cases where we already have results for codeML [4] and OmegaMap [8], and there is evidence for recombination using the full sequence dataset (amino acid datasets for Zika genes C, E, NS1, NS2A, and NS3). For genes E, NS1, NS2A and NS3, no positively selected amino acid sites were detected, in agreement with the codeML and OmegaMap results. For the C coding region, a single positively selected amino acid site was identified by FUBAR at position 8 of the alignment. This position is inferred to be positively selected with a probability of 0.84 and 0.94 by codeML model M8 (Naive Empirical Bayes and Bayes Empirical Bayes analyses, respectively), and by OmegaMap with a probability of 0.75. We do not consider this as a discrepancy but rather a sensitivity issue related with the small number of sequences (15) being analysed. Therefore, we conclude that FUBAR is not too sensitive to recombination.

Given the above results, we propose as the most efficient and fastest way of analysing a large amount of viral coding sequences, the protocol depicted in Fig. 1. SEDA [11] is used for data preparation and filtering. The first step of the protocol is a fast screen for positive selection, using the clustalomega [17] alignment algorithm, FastTree [18] for phylogenetic inferences, and FUBAR [5] for the detection of positively selected amino acid sites, with an estimated running time of a few minutes only, even when using a large number of sequences. This is a fast way of identifying genes that likely shows signs of positive selection and thus should be further analysed. The second step is a more time-consuming reevaluation of those genes lacking evidence for positive selection after the first step, using clustalomega, FastTree and, for positive selection inferences, codeML [4] models M1a and M2a (that could take a couple of days to run, depending on the value of *p*). The third step is a highly time consuming and accurate analysis of the genes that show evidence for positive selection after the first two steps, using Muscle [14] for sequence alignment, MrBayes [15] for phylogenetic inferences, and for the detection of positively selected amino acid sites FUBAR and codeML models M7 and M8 (that could take several days to run, depending on the value of *p*). If recombination is detected in the dataset when using the Phitest (as implemented in PhiPack (see Footnote 6)), then we also run the insensitive to recombination population-based method OmegaMap [8] for the identification of positively selected amino acid sites. The positively selected amino acid sites that are identified by more than one method are likely true. Because of the limitation of codeML regarding *p*, if not all putative positively selected amino acid sites are identified by all methods, up to four additional replicates (data allowing) are taken and analysed the same way. This is why when preparing the data for the third step, up to 1000 randomly selected sequences that are different at the amino acid level are first aligned using Muscle, and then 200 sequences are randomly taken from this alignment.

4 Conclusion

A protocol is proposed to analyze a large number of datasets, such as those available at the ViPR database. In the future, we aim at including OmegaMap, FUBAR and PhiPack in ADOPS and adjust accordingly the B+ layout.

Acknowledgments. This article is a result of the project Norte-01-0145-FEDER-000008 - Porto Neurosciences and Neurologic Disease Research Initiative at I3S, supported by Norte Portugal Regional Operational Programme (NORTE 2020), under the PORTUGAL 2020 Partnership Agreement, through the European Regional Development Fund (FEDER). The SING group thanks the CITI (Centro de Investigación, Transferencia e Innovación) from the University of Vigo for hosting its IT infrastructure. This work was partially supported by the Consellería de Educación, Universidades e Formación Profesional (Xunta de Galicia) under the scope of the strategic funding ED431C2018/55-GRC Competitive Reference Group. H. López-Fernández is supported by a post-doctoral fellowship from Xunta de Galicia (ED481B 2016/068-0).

References

1. Twiddy, S.S., Woelk, C.H., Holmes, E.C.: Phylogenetic evidence for adaptive evolution of dengue viruses in nature. J. Gen. Virol. **83**, 1679–1689 (2002). https://doi.org/10.1099/0022-1317-83-7-1679
2. Woelk, C.H., Holmes, E.C.: Variable immune-driven natural selection in the attachment (G) glycoprotein of respiratory syncytial virus (RSV). J. Mol. Evol. **52**, 182–192 (2001)

3. Woelk, C.H., Jin, L., Holmes, E.C., Brown, D.W.G.: Immune and artificial selection in the haemagglutinin (H) glycoprotein of measles virus. J. Gen. Virol. **82**, 2463–2474 (2001). https://doi.org/10.1099/0022-1317-82-10-2463
4. Yang, Z.H.: PAML 4: Phylogenetic analysis by maximum likelihood. Mol. Biol. Evol. **24**, 1586–1591 (2007). https://doi.org/10.1093/molbev/msm088
5. Murrell, B., Moola, S., Mabona, A., Weighill, T., Sheward, D., Kosakovsky-Pond, S.L., Scheffler, K.: FUBAR: a fast, unconstrained bayesian approximation for inferring selection. Mol. Biol. Evol. **30**, 1196–1205 (2013). https://doi.org/10.1093/molbev/mst030
6. Shriner, D., Nickle, D.C., Jensen, M.A., Mullins, J.I.: Potential impact of recombination on sitewise approaches for detecting positive natural selection. Genet. Res. **81**, 115–121 (2003)
7. Anisimova, M., Nielsen, R., Yang, Z.H.: Effect of recombination on the accuracy of the likelihood method for detecting positive selection at amino acid sites. Genetics **164**, 1229–1236 (2003)
8. Wilson, D.J., McVean, G.: Estimating diversifying selection and functional constraint in the presence of recombination. Genetics **172**, 1411–1425 (2006). https://doi.org/10.1534/genetics.105.044917
9. Reboiro-Jato, D., Reboiro-Jato, M., Fdez-Riverola, F., Vieira, C.P., Fonseca, N.A., Vieira, J.: ADOPS–Automatic Detection Of Positively Selected Sites. J. Integr. Bioinform. **9**, 200 (2012). https://doi.org/10.2390/biecoll-jib-2012-200
10. Vázquez, N., Vieira, C.P., Amorim, B.S.R., Torres, A., López-Fernández, H., Fdez-Riverola, F., Sousa, J.L.R., Reboiro-Jato, M., Vieira, J.: Large scale analyses and visualization of adaptive amino acid changes projects. Interdiscip. Sci. **10**, 24–32 (2018). https://doi.org/10.1007/s12539-018-0282-7
11. López-Fernández, H., Duque, P., Henriques, S., Vázquez, N., Fdez-Riverola, F., Vieira, C. P., Reboiro-Jato, M., Vieira, J.: Bioinformatics protocols for quickly obtaining large-scale data sets for phylogenetic inferences. Interdiscip. Sci. **11**, 1–9 (2019). https://doi.org/10.1007/s12539-018-0312-5
12. Bruen, T.C., Philippe, H., Bryant, D.: A simple and robust statistical test for detecting the presence of recombination. Genetics **172**, 2665–2681 (2006). https://doi.org/10.1534/genetics.105.048975
13. Kloepper, T.H., Huson, D.H.: Drawing explicit phylogenetic networks and their integration into SplitsTree. BMC Evol. Biol. **8**, 22 (2008). https://doi.org/10.1186/1471-2148-8-22
14. Edgar, R.C.: MUSCLE: multiple sequence alignment with high accuracy and high throughput. Nucleic Acids Res. **32**, 1792–1797 (2004). https://doi.org/10.1093/nar/gkh340
15. Ronquist, F., Huelsenbeck, J.P.: MrBayes 3: Bayesian phylogenetic inference under mixed models. Bioinformatics **19**, 1572–1574 (2003)
16. Weaver, S., Shank, S.D., Spielman, S.J., Li, M., Muse, S.V., Pond, S.L.K.: Datamonkey 2.0: a modern web application for characterizing selective and other evolutionary processes. Mol. Biol. Evol. **35**, 773–777 (2018). https://doi.org/10.1093/molbev/msx335
17. Sievers, F., Higgins, D.G.: Clustal Omega for making accurate alignments of many protein sequences. Protein Sci. **27**, 135–145 (2018). https://doi.org/10.1002/pro.3290
18. Price, M.N., Dehal, P.S., Arkin, A.P.: FastTree 2-approximately maximum-likelihood trees for large alignments. PLoS ONE **5**, e9490 (2010). https://doi.org/10.1371/journal.pone.0009490
19. Liu, K., Linder, C.R., Warnow, T.: RAxML and FastTree: comparing two methods for large-scale maximum likelihood phylogeny estimation. PLoS ONE **6**, e27731 (2011). https://doi.org/10.1371/journal.pone.0027731

Data-Driven Extraction of Quantitative Multi-dimensional Associations of Cardiovascular Drugs and Adverse Drug Reactions

Upasana Chutia, Jerry W. Sangma, Vipin Pal(✉), and Yogita(✉)

National Institute of Technology Meghalaya, Shillong, India
upasananitm@gmail.com, jaywatsan@gmail.com, vipinrwr@gmail.com, thakranyogita@gmail.com

Abstract. Early detection of adverse drug reactions as a part of post-marketing surveillance is very crucial for saving a number of persons from unwanted consequences of drugs. Along with the drug, patient's traits such as age, gender, weight, location are key factors for occurrence of adverse effects. The relationship between drug, patient attributes and adverse drug effects can be precisely represented by quantitative multi-dimensional association rules. But discovery of such rules faces the challenge of data sparsity because of the large number of possible side effects of a drug and fewer number of corresponding data records. In this paper, to address the data sparsity issue, we propose to use variable support based LPMiner technique for detecting quantitative multi-dimensional association rules. For experimental analysis, data corresponding to three cardiovascular drugs namely Rivaroxaban, Ranolazine and Alteplase has been taken from U.S. FDA Adverse Event Reporting System database. The experimental results show that based on LPMiner technique a number of association rules have been detected which went undetected in case of constant support based apriori and FP-Growth technique.

Keywords: Multi-dimensional associations rules · Pharmacovigilance · Data mining · Adverse Drug Reactions (ADR)

1 Introduction

Any injury or side effect occurring in a patient on taking normal dose of a drug is termed as Adverse Drug Reaction (ADR) [1]. Drugs are prescribed to a number of patients for betterment of their health and if any drug causes an adverse effect then it may result in affecting a large population. Sometime these ADR can be life threatening and can lead to permanent disability of any body part or even death in some cases [2]. Timely detection of ADR is very necessary so that early warning signals can be raised against these adverse reactions to save a large population from injurious consequences of drugs intake [3]. Detection,

F. Fdez-Riverola et al. (Eds.): PACBB 2019, AISC 1005, pp. 70–77, 2020.
https://doi.org/10.1007/978-3-030-23873-5_9

monitoring and prevention of ADR related to medical drugs fall under Pharmacovigilance [4–6]. Generally two approaches are followed for detection of ADR namely Pre-marketing clinical trials and Post-marketing surveillance. Clinical trials are conducted before licensing a drug to market for analyzing its suitability on a sample population over a short period of time. Various adverse effects of drugs remain unidentified in clinical trials. Post-marketing surveillance aims to detect those adverse effects of licensed drugs which have not been accounted in pre-marketing clinical trials. For this purpose, data is collected in the form of adverse effects reports which are submitted by consumers, pharmacists, physicians and other healthcare professionals [7]. In several countries online systems have been made operational for spontaneous, easy and continual collection of such reports. Nowadays, tremendous volume of data is getting collecting through these systems which contain very useful information related to drugs and ADR, hidden into it. Extracting this information manually is very tedious and difficult.

Data mining deals with the extraction of useful patterns from large amount of data [8]. Association rule mining is a data mining approach which has been used in a number of research works for finding associations between drugs and ADR [9–12]. Along with the drug, these are the patient's traits such as gender, age, weight, place of living, which play a vital role in occurrence of an ADR. So identifying associations between drugs, patient's traits and adverse drug effects is very important. This type of relationships can be precisely represented by the Quantitative Multi-dimensional Association Rules (QMAR) (for details of QMAR, refer to Sect. 2). But most of association rule mining methods face the challenge of data sparsity whenever it comes to detection of QMAR. Data sparsity signifies a situation where the possible ADR for a drug are many and corresponding data examples are few. Apriori [13] and FP-Growth [14] methods are two association rule mining methods that have been highly referenced for finding association between drugs and ADR. These methods consider a constant threshold for support of association rules irrespective of dimensionality of rules. But this assumption is not suitable in case of high dimensional and sparse data hence a number of multi-dimensional rules left undetected in case of such methods.

In this paper, we propose to employ variable support based technique namely Long Pattern Miner (LPMiner) [15] which decreases the support threshold as the dimensionality of frequent itemsets increases for finding QMAR. The performance of LPMiner has been compared with Apriori and FP-Growth on the dataset of three cardiovascular drugs viz. Rivaroxaban, Ranolazine and Alteplase that has been taken from U.S. FDA Adverse Event Reporting System (FAERS) database [7]. The experimental results show that LPMiner has outperformed other two methods in terms of detected QMAR.

The rest of this paper is arranged as follows: in the next section, preliminary details of association rule mining have been described briefly. Dataset details have been provided in Sect. 3. In Sect. 4, we discuss the proposed methodology. Analysis of experimental results has been presented in Sect. 5. The conclusion of this paper and future work has been given in Sect. 6.

2 Preliminaries

2.1 Association Rule

Given a set I = {$x_1, x_2, \ldots x_n$} of n data attributes called items and a dataset D = $r_1, r_2, \ldots r_m$} containing m data records where each record r_i comprises subset of the items in I then a association rule is defined as an implication of the form $X \Rightarrow Y$ where $X \subset I$, $Y \subset I$, $X \neq \phi$, $Y \neq \phi$ and $X \cap Y = \phi$. Each association rule is made up of two different sets of data items, called itemsets, X and Y which are respectively called antecedent and consequent of rule [8,16]. For an association rule to be useful, it must satisfy some rule metrics. Support, confidence and lift, defined as follows, are well known evaluation metrics for association rules [8,16].

- **Support** ($X \Rightarrow$ **Y**) - It is defined as the ratio of the number of records that contain both X and Y to the total number of records in the dataset.
- **Confidence** ($X \Rightarrow$ **Y**) - It is defined as the ratio of the number of records that contain both X and Y to the number of records that contain X irrespective of Y as given in Eq. (1) .

$$Confidence(X \Rightarrow Y) = \frac{(Support(X \cup Y))}{(Support(X))} \tag{1}$$

- **Lift** - It is defined as the ratio of the observed support of the rule to the expected support if X and Y were independent as given in Eq. (2).

$$Lift(X \Rightarrow Y) = \frac{(Support(X \cup Y))}{(Support(X) \times Support(Y))} \tag{2}$$

  It is a measure that takes statistical dependence into consideration. If $lift = 1$, it signifies that occurrences of X and Y are independent of each other. If $lift > 1$, it signifies that X and Y are positively correlated. If $lift < 1$, it represent that there is a negative correlation between X and Y.

2.2 Quantitative Multi-dimensional Association Rule (QMAR)

It is an association rule which contains two or more than two dimensions and each dimension have a specific value or range of values associated with it [8,16]. Here, the dimensions represent data attributes of records. Data attributes can be categorical or numeric. An example of QMAR is as follows:

$$Age = 15\text{–}25 \wedge Weight = 80\text{–}100 \Rightarrow Medical\ Condition = Obesity$$

2.3 Association Rule Mining

It is a process of extracting interesting association rules from the data which represent the correlations, associations, frequent patterns or causal relations prevailing among sets of data items in a given dataset. Association rule mining is a two-step process [16].

Step 1: Find all itemsets that have a support higher than a given threshold which are called frequent itemsets. It can be done using Apriori [13] or FP-Growth [14] or LPMiner [15] or any other frequent pattern mining algorithm. Providing details of these algorithms is out of the scope of this paper. For details please refer to [13–15].
Step 2: Generate rules for each itemset as follows:
Given a frequent itemset F, find all non-empty subsets $X \subset F$ such that $X \Rightarrow F - X$, if *confidence*$(X \Rightarrow F - X)$ is higher than minimum confidence threshold then $X \Rightarrow F - X$ is valid and useful. Similarly generate valid rules corresponding to all frequent itemsets discovered in Step 1.

2.4 LPMiner Algorithm

The main property of most of the algorithms for finding frequent itemsets is that they use a constant support threshold irrespective of the size of itemsets. Itemsets that comprise only a few items tend to be valid and useful as they tend to have a high support. Long itemsets can be valid and useful even though their support is relatively small. To find small as well as long itemsets, LPMiner considers variable support threshold which decreases as a function of itemsets length [15]. As LPMiner is capable to detect long frequent itemsets it frames a base for finding QMAR.

3 Dataset

Dataset for the present work has been taken from the FDA Adverse Event Reporting System. It maintains a database of adverse event reports submitted by pharmaceutical companies, pharmacists, physicians, medical professionals and consumers as a post-marketing surveillance activity for drugs. This database is hosted on U.S. Food and drug administration's website [7]. Dataset of three cardiovascular drugs namely Rivaroxaban, Ranolazine and Alteplase and related ADR has been filtered by selecting cardiovascular, cardiac arrest, cardiac disease, heart attack, heart failure terms under indication attribute of FAERS database for each of the three drugs. Indications attribute represents the disease for which drug has been prescribed. Total number of records retrieved over a period of 5 years from July 2013 to June 2018 for each drug are given in Table 1.

Table 1. Dataset

Drug name	Count of records
Rivaroxaban	3374
Ranolazine	2239
Alteplase	1837

Each record comprises data values corresponding to 6 attributes viz. Drug, Age, Weight, Gender, Country Code and ADR. Data values for age attribute have different units such as days, weeks, months and decades; all such values are mapped to years. Similarly, weight attribute has data values in pounds and grams which are converted into kilograms.

4 Methodology

The proposed methodology for finding QMAR for cardiovascular drugs comprises of three steps:

1. Data Preprocessing
2. Finding Frequent Itemsets
3. Association Rule Generation

4.1 Data Preprocessing

- **Impute Missing Values** - There are several missing values in the dataset corresponding to age, weight, gender and country code. A number of research works that have processed AERS data have deleted the records having missing values but in the present work instead of deleting such records, missing values are imputed in those records by applying k-NN [18] approach. Because removal of such records results in information loss which degrades quality of results.
- **Data Transformation** - Firstly, continuous attributes such as age and weight are discretized into different categories. Same categories have been created as given in [10,17]. Further, categorical attributes such as gender, country code, ADR, including discretized age and weight, are converted to binomial form. This conversion is done by creating a binary attribute corresponding to each attribute-value pair of a attribute for all above mentioned attributes.

4.2 Finding Frequent Itemsets

LPMiner algorithm is applied to preprocessed data for finding frequent itemsets. Initially, the minimum support threshold is fixed at 0.1 for two-dimensional itemsets and then decreased by 50% whenever itemsets length increases by one dimension. The reason for keeping support threshold low is that data examples corresponding to each drug and ADR pair are fewer in comparison to total dataset size. This support threshold can be justified in the light of fact that we are focusing on a real world problem falling under medical domain where collecting large number of real examples corresponding to a particular drug and ADR pair is very difficult. But extracting QMAR for drugs and ADR is really very important because even a value of 0.1 support threshold represents that approximately 337, 223 and 183 patients corresponding to Rivaroxaban, Ranolazine and Alteplase, out of a total of 3374, 2239 and 1837 patients respectively faced at least one side effect of a particular drug.

4.3 Association Rule Generation

A number of frequent itemsets are generated by LPMiner but all of them are not required so to discard irrelevant frequent itemsets and generate QMAR the following steps are followed:

Step 1: Out of the frequent itemsets generated by LPMiner discards such itemsets that do not comprise both drug and ADR as their member item.
Step 2: Some of the remaining frequent itemsets may lead to generation of such association rules that have drug as rule consequent or ADR as rule antecedent or patient traits as consequent but such rules are not representative of associations between drugs and ADR properly. To avoid generation of such rules a constraint based rule generation approach is followed whereby only those association rules are produced that have structure as per a rule template [14]. For the purpose of generating only useful QMAR following template has been considered:

$$\textit{Drug} \wedge \textit{patient trait (1)} \wedge ..... \wedge \textit{patient trait(k)} \Rightarrow \textit{ADR} \{\textit{Confidence} \geq \textit{0.4}\}$$

where drug can be any drug out of Rivaroxaban, Ranolazine and Alteplase and ADR is any one out of all possible ADR of respective drug. For a rule to be valid confidence should be greater than or equal to 0.4.
Step 3: Compute lift for all QMAR produced in Step 2.

5 Experimental Results and Analysis

The total number of QMAR detected for three cardiovascular drugs Rivaroxaban, Ranolazine and Alteplase based on LPMiner, Apriori and FP-Growth techniques are shown in Table 2. Samples of QMAR detected corresponding to each of the drug are given in Table 3.

Table 2. Total number of QMAR detected by LPMiner, Apriori and FP-Growth for three cardiovascular drugs

QMAR discovery technique	Number of quantitative multi-dimensional association rules (QMAR)				
	2-Dimensional	3-Dimensional	4-Dimensional	5-Dimensional	6-Dimensional
LPMiner	06	09	41	03	00
Apriori	06	03	00	00	00
FP-Growth	06	03	00	00	00

It can be observed from Table 2 that in total, 59 QMAR have been discovered based on LPMiner method, out of which 06 are two-dimensional, 09 are three-dimensional, 41 are four-dimensional and 03 are five-dimensional whereas Apriori and FP-Growth have detected total 09 QMAR, out of which 06 are

Table 3. Sample QMAR extracted for Rivaroxaban, Ranolazine and Alteplase

Drug name	Quantitative multi-dimensional association rules (QMAR)
Rivaroxaban	Rivaroxaban ∧ Female ∧ Age(0–6) ⇒ Dyspnoea { Lift(40.71)}
	Rivaroxaban ∧ Female ∧ Age(44–64) ⇒ Gastrointestinal Hemorrhage { Lift(5.99)}
Ranolazine	Ranolazine ∧ Weight(101–120) ∧ Female ⇒ Death {Lift(12.53)}
	Ranolazine ∧ Female ∧ Age(44–64) ⇒ Myocardial Infarction {Lift(21.5)}
Alteplase	Alteplase ∧ Weight(81–100) ∧ Female ⇒ Death {Lift(4.65)}
	Alteplase ∧ Weight(31–50) ∧ Female ⇒ Neurological Decompensation {Lift(18.1)}
	Alteplase ∧ Weight(81–100) ∧ Age(44–64) ⇒ Hemorrhagic Transformation Stroke {Lift(8.83)}

two-dimensional and 03 are three-dimensional. It shows that the performance of LPMiner is better than the performance of other two methods in terms of QMAR detection. In totality, it can be said that LPMiner is capable to detect QMAR as it varies support threshold as per the length of itemsets as opposed to other methods which keep constant support threshold. Because of this many multidimensional itemsets do not satisfy the support threshold and a number of important QMAR go undetected.

It can be seen from Table 3 that drug rivaroxaban has dyspnoea and gastrointestinal hemorrhage as its associated adverse effects in case of female infants and middle aged females respectively. Similarly, drug Ranolazine has death as its side effect in case of overweight females and myocardial infarction if patient is a middle age woman. Alteplase associated adverse reactions are death, neurological decompensation and hemorrhagic transformation stroke for females considering their age and weight. For most of the detected QMAR, we found that lift value is quite high which implies that these rules are statistically important and not occurring by chance. This type of information related to drugs and ADR can be very helpful for physicians in framing treatment process and in reducing harmful consequences of drugs.

6 Conclusion and Future Work

In this paper, a methodology for finding QMAR of cardiovascular drugs and patient traits with drug adverse reactions from U.S. FDA AERS dataset has been presented. For this purpose, a data mining technique LPMiner has been applied. On comparing results of LPMiner with other techniques viz. Apriori and FP-Growth, it is concluded that for LPMiner higher number of associations have been detected. In case of LPMiner, total 59 QMAR have been extracted whereas for Apriori and FP-Growth total 09 QMAR for each of the two have been extracted. In future, as an extension of this work, we will focus on identifying exceptional associations based on high dimensional data mining approaches.

References

1. George, C.: Reporting Adverse Drug Reactions: A Guide for Healthcare Professionals. British Medical Association, London (2006)

2. Rockville, M.D.: Reducing and Preventing Adverse Drug Events To Decrease Hospital Costs: Research in Action. Agency for Healthcare Research and Quality, Issue 1, March 2014
3. Chen, Y., Guo, J.J., Steinbuch, M.: Comparison of sensitivity and timing of early signal detection of four frequently used signal detection methods: an empirical study based on the US FDA adverse event reporting system database. Pharm Med. **22**, 359–365 (2008)
4. World Health Organization (WHO): The Importance of Pharmacovigilance: Safety Monitoring of Medicinal Products (2002)
5. Lindquist, M.: The need for definitions in pharmacovigilance. Drug Saf. **30**, 825–830 (2007)
6. World Health Organization: The Importance of Pharmacovigilance. WHO Collaborating Centre for International Drug Monitoring, Geneva, vol. 44 (2002)
7. U.S. Food and Drug Administration, FAERS quarterly data files. https://www.fda.gov/drugs/guidancecomplianceregulatoryinformation/surveillance/adversedrugeffects/default.Htm. Accessed Oct 2018
8. Han, J., Kamber, M.: Data Mining: Concepts and Techniques. Elsevier, Amsterdam (2011)
9. Wang, C., et al.: Exploration of the association rules mining technique for the signal detection of adverse drug events in spontaneous reporting systems. PloS One **7**(7) (2012)
10. Yildirim, P., Ilyas, O., Holzinger, A.: On knowledge discovery in open medical data on the example of the FDA drug adverse event reporting system for alendronate (FOSAMAX). In: Holzinger, A., Pasi, G. (eds.) Human-Computer Interaction and Knowledge Discovery in Complex, Unstructured, Big Data, pp. 195–206. Springer, Heidelberg (2013)
11. Reps, J.M., Aickelin, U., Ma, J., Zhang, Y.: Refining adverse drug reactions using association rule mining for electronic healthcare data. In: Proceedings of IEEE International Conference on Data Mining Workshop (ICDMW 2014), pp. 763–770 (2014)
12. Yang, X., Albin, A., Ren, K., Zhang, P., Etter, J.P., Lin, S., Li, L.: Efficiently mining adverse event reporting system for multiple drug interactions. AMIA Summits Transl. Sci. Proc. **120** (2014)
13. Agrawal, R., Srikant, R.: Fast algorithms for mining association rules. In: Proceedings of 20th International Conference on Very Large Data Bases, VLDB, vol. 1215 (1994)
14. Han, J., Pei, J., Yin, Y.: Mining frequent patterns without candidate generation. In: ACM SIGMOD Record, vol. 29, no. 2. ACM (2000)
15. Seno, M., Karypis, G.: LPMiner: an algorithm for finding frequent itemsets using length-decreasing support constraint. In: Proceedings of IEEE International Conference on Data Mining (ICDM 2001) (2001)
16. Tan, P.-N.: Introduction to Data Mining. Pearson Education India, New Delhi (2007)
17. Sakaeda, T., Kadoyama, K., Okuno, Y.: Adverse event profiles of platinum agents: data mining of the public version of the FDA adverse event reporting system, AERS, and reproducibility of clinical observations. Int. J. Med. Sci. **8**(6), 487–491 (2011)
18. Dixon, J.K.: Pattern recognition with partly missing data. IEEE Trans. Syst. Man Cybern. **9**, 617–621 (1979)

An Identical String Motif Finding Algorithm Through Dynamic Programming

Abdelmenem S. Elgabry(✉), Tahani M. Allam(✉), and Mahmoud M. Fahmy(✉)

Faculty of Engineering, Tanta University, Tanta, Egypt
abdo55@gmail.com, tahany@f-eng.tanta.edu.eg, mfn_288@hotmail.com

Abstract. Gene expression regulation is a major challenge in biology. One aspect of such a challenge is the binding sites in DNA, called motifs. DNA motif finding still poses a great challenge for computer scientists and biologists. As a result, a large number of motif finding algorithms are already implemented. However, literature has proven this task to be complex. The present paper tends to find a solution for the motif finding problem through rearranging data in a manner that can help obtain the targeted motif easily by adopting the dynamic programming concept. It proposes an efficient algorithm called Pattern Position Motif Finding (PPMF), aiming at finding all identical string motifs, which appear in a single sequence or multi sequences at least twice or a specified times. The proposed algorithm is compared with the Encoded Expansion (EE) algorithm to evaluate the execution time and size of processed sequences, PPMF takes less execution time than the corresponding one and processed large size sequences than EE processed. This denotes that when the biologist needs to find the identical string motifs in a big sequence, our proposed algorithm will be the better solution than the EE algorithm.

Keywords: Identical string motifs · Gene expression regulation · DNA motifs · Nucleotide and protein sequences · Sequence analysis

1 Introduction

A DNA motif is a nucleotide or amino acid sequence pattern, has been acknowledged to have a special biological significance [1]. Motifs represent the repeated patterns presenting DNA or protein. This segment of nucleotide or amino acid segments are arranged in a specific structure which acts as a signature for a particular family. These motifs are considered a signature structure of the protein identifying its family.

Motif discovery has important roles in the area of locating regularity sites and drug target identification in biological sequences. Regulatory sites on a DNA sequence normally correspond to shared conservative sequence patterns among the regulatory regions of correlated genes [2]. Identifying motifs and the corresponding instances is very important, so biologists can investigate the interactions between DNA and proteins, gene regulation, cell development and cell reaction under physiological and pathological conditions [3]. Biologists and computer scientists face the difficulty of

F. Fdez-Riverola et al. (Eds.): PACBB 2019, AISC 1005, pp. 78–86, 2020.
https://doi.org/10.1007/978-3-030-23873-5_10

DNA motifs. As such, motif-finding algorithms have been implemented [4], Karci [5] and Encoded Expansion (EE) [6, 7]. Karci has proposed a deterministic algorithm via which he gets all identical string motifs that appear twice or more times with all possible sizes. Yet, literature has proven this algorithm inefficient in time and space [7]. The compared EE algorithm gives more improvement over Karci's, allowing for setting the minimum number of motifs.

In general, motif finding problem has, in essence, to do with multiple sequence alignment, a problem that accounts for the present paper's focus: introducing an algorithm which does not align with the sequence. It depends on dynamic programming techniques, and finds all motifs appearing in a single sequence. It deals with multi-sequences in their totality as a single sequence after concatenating them with a separator of a special character "$" as used herein. An identical string motif can be either the exact pattern in all sequences or more than one pattern in a single sequence. The present algorithm adopts the methodology followed by the EE algorithm and has been improved by ignoring the coding and decoding of sequence nucleotides. It also depends on direct pattern positioning. The principle of the algorithm methodology will be dealt with below.

The issue is given a sequence or sequences, find the motifs present in the sequence or in all sequences.

2 Methodology

The proposed algorithm adopts a dynamic programming strategy in order to develop an efficient motif-finding algorithm. Dynamic programming is a powerful technique to solve problems that can be split up into some "overlapping" smaller sub-problems. It starts solving from the trivial sub-problem, up toward the given problem. The proposed algorithm goes over several reinforcements in an attempt to improve the overall performance.

The sample sequence: ACTCAGCTACCTCAGTACACTCAG is used to illustrate the reinforcements of the proposed algorithm. Figure 1 shows the index of each nucleotide in the sample sequence.

position	0	1	2	3	4	5	6	7	8	9	10	11	12	13	14	15	16	17	18	19	20	21	22	23
Sequence	A	C	T	C	A	G	C	T	A	C	C	T	C	A	G	T	A	C	A	C	T	C	A	G

Fig. 1. The index of the sample sequence.

Initially, the algorithm scans the given sample sequence in order to record the occurrence positions of all the 2 bases sub-sequences and the next nucleotide (Fig. 2(a)).

Only the subsequences that appear at least twice (or the minimum number of required occurrences) are considered 2 bases motifs, referred to as 2k motifs, while other sequences are neglected (Fig. 2(b)).

2K patterns	positions	next nucleotide
AA	[ ]	
AC	[0, 8, 16, 18]	[T, C, A, T]
AG	[4, 13, 22]	[C, T, -]
AT	[ ]	
CA	[3, 12, 17, 21]	[G, G, C, G]
CC	[9]	[T]
CG	[ ]	
CT	[1, 6, 10, 19]	[C, A, C, C]

2K patterns	positions	next nucleotide
GA	[ ]	
GC	[5]	[T]
GG	[ ]	
GT	[14]	[A]
TA	[7, 15]	[C, C]
TC	[2, 11, 20]	[A, A, A]
TG	[ ]	
TT	[ ]	

(a)

2K Motifs	positions	next nucleotide
AC	[0, 8, 16, 18]	[T, C, A, T]
AG	[4, 13, 22]	[C, T, -]
CA	[3, 12, 17, 21]	[G, G, C, G]
CT	[1, 6, 10, 19]	[C, A, C, C]
TA	[7, 15]	[C, C]
TC	[2, 11, 20]	[A, A, A]

(b)

Fig. 2. **(a)** The specified positions and next nucleotide of all 2 bases sub sequences. **(b)** The specified positions of all 2k motifs.

In the next stage, the 2k occurrence positions and the next nucleotide come under observation to identify the consecutive sub-sequences, extracting the 3k as shown in (Fig. 3(a)). If a nucleotide appears more than once in the next nucleotide columns, this means that the corresponding 2k motif patterns with this nucleotide are repeated in the sequence by the number of appearance of this nucleotide. Once the nucleotide appears at least twice (or the minimum number of required occurrences), it shall be combined with the corresponding 2k motif pattern to get the 3k motifs. The next nucleotide 'T', for example, appears twice after the 'AC' pattern, accounting for the pattern 'ACT' appearance at the positions 0 and 18. The 'ACT' pattern will be considered a motif, because it appears twice as shown in (Fig. 3(a)).

The nucleotides 'C' and 'A' appear only once in the next nucleotide column for the 'AC' pattern, which means that the 'ACC' and the 'ACA' patterns are not repeated in the sample sequence as shown in (Fig. 3(a)), The 'ACC' and 'ACA' patterns are not considered a motif. Using the same strategy, all the 3k motifs can be obtained. After getting each 3k motif and its occurrence positions (Fig. 3(b)), the result is to be analyzed so that the next nucleotide of all 3k motifs is determined. The new data, including the positions of the 3k motifs and the next nucleotide, is used to extract the 4k motifs (Fig. 4). Ultimately, a new result presenting the 4k motifs and their corresponding positions is at hand (Fig. 4(b)). Using the same strategy, we can get all the K motifs.

To enhance the performance of the proposed algorithm, a modification on the methodology is to be undertaken; a minor update is done to the algorithm. Instead of extracting 3k motifs from the 2k motifs, scanning the given sample sequence to record the occurrence positions of all the 3 bases subsequence is identical to what is done with the 2k motifs, which will improve efficiency and increase the computation speed. However, this is infusible when the size of the subsequence exceeds 3 bases, because indexing k bases subsequence is an exponential problem of order 4^k.

In this update, the algorithm scans the given sample sequence in order to record the occurrence positions of all the 2 bases subsequences only; there is no need to record the next nucleotide here. The patterns that do not satisfy the motifs conditions are to be neglected, while the remaining patterns are selected to get the 2k motif. The next nucleotide will not be recorded as in the previous technique see (Fig. 5). After that, the algorithm rescans the given sample sequence in order to record the occurrence positions of all the 3 bases sub sequences and the next nucleotide (Fig. 6).

2K Motifs	positions	next nucleotide
AC	[0, 8, 16, 18]	[T, C, A, T]
AG	[4, 13, 22]	[C, T, -] ✗
CA	[3, 12, 17, 21]	[G, G, C, G]
CT	[1, 6, 10, 19]	[C, A, C, C]
TA	[7, 15]	[C, C]
TC	[2, 11, 20]	[A, A, A]

(a)

3k

3 K Motifs	positions
ACT	[0, 18]
CAG	[3, 12, 21]
CTC	[1, 10, 19]
TAC	[7, 15]
TCA	[2, 11, 20]

(b)

Fig. 3. Shows the 2k motifs as well as their specified positions arranged and next nucleotide and how to get the 3k motifs from it.

3 K Motifs	positions		next nucleotide
ACT	[0, 18]	read next nucleotide at position + 3 for each position in 3k motifs positions	[C, C]
CAG	[3, 12, 21]		[C, T, -]
CTC	[1, 10, 19]		[A, A, A]
TAC	[7, 15]		[C, A]
TCA	[2, 11, 20]		[G, G, G]

(a)

4k

4 K Motifs	positions
ACTC	[0, 18]
CTCA	[1, 10, 19]
TCAG	[2, 11, 20]

(b)

Fig. 4. Extracting the 4k motifs from the 3k motifs.

2k patterns	positions	patterns	positions
AA	[ ] ✗	GA	[ ] ✗
AC	[0, 8, 16, 18]	GC	[5] ✗
AG	[4, 13, 22]	GG	[ ] ✗
AT	[ ] ✗	GT	[14] ✗
CA	[3, 12, 17, 21]	TA	[7, 15]
CC	[9] ✗	TC	[2, 11, 20]
CG	[ ] ✗	TG	[ ] ✗
CT	[1, 6, 10, 19]	TT	[ ] ✗

(a)

2K

2K Motifs	positions
AC	[0, 8, 16, 18]
AG	[4, 13, 22]
CA	[3, 12, 17, 21]
CT	[1, 6, 10, 19]
TA	[7, 15]
TC	[2, 11, 20]

(b)

Fig. 5. **(a)** The 2k patterns and their specified positions. **(b)** The 2k motifs.

Then, the patterns satisfying the motifs condition are selected while neglecting the other patterns as shown in (Fig. 7(a)). The 4k motifs will be extracted from the 3k motifs after scanning the next nucleotide for resulted 3k motifs see (Fig. 7(b)) and the 5k motifs will be extracted from the 4k motifs. Given that, the 3k pattern needs an array up to 4^3 which equals 64 elements; the length of processed data accounts for this small array.

3K patterns	positions	next nucleotide	3K patterns	positions	next nucleotide	3K patterns	positions	next nucleotide	3K patterns	positions	next nucleotide
AAA	✗		CAA	✗		GAA	✗		TAA	✗	
AAC	✗		CAC	[17] ✗	[T]	GAC	✗		TAC	[7, 15]	[C, A]
AAG	✗		CAG	[3, 12, 21]	[C, T, -]	GAG	✗		TAG	✗	
AAT	✗		CAT	✗		GAT	✗		TAT	✗	
ACA	[16] ✗	[C]	CCA	✗		GCA	✗		TCA	[2, 11, 20]	[G, G, G]
ACC	[8] ✗	[T]	CCC	✗		GCC	✗		TCC	✗	
ACG	✗		CCG	✗		GCG	✗		TCG	✗	
ACT	[0, 18]	[C, C]	CCT	[9] ✗	[C]	GCT	[5] ✗	[A]	TCT	✗	
AGA	✗		CGA	✗		GGA	✗		TGA	✗	
AGC	[4] ✗	[T]	CGC	✗		GGC	✗		TGC	✗	
AGG	✗		CGG	✗		GGG	✗		TGG	✗	
AGT	[13] ✗	[A]	CGT	✗		GGT	✗		TGT	✗	
ATA	✗		CTA	[6] ✗	[C]	GTA	[14] ✗	[A]	TTA	✗	
ATC	✗		CTC	[1, 10, 19]	[A, A, A]	GTC	✗		TTC	✗	
ATG	✗		CTG	✗		GTG	✗		TTG	✗	
ATT	✗		CTT	✗		GTT	✗		TTT	✗	

Fig. 6. The specified positions and next nucleotide of all 3 bases sub sequences.

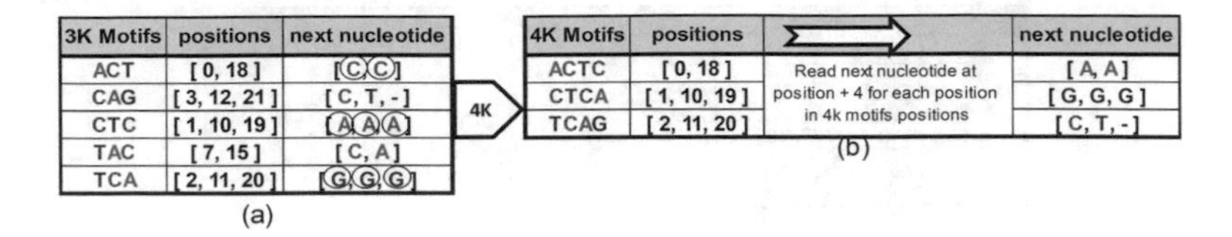

Fig. 7. Shows the 4k motifs as well as their specified positions and next nucleotide.

3 Algorithm Implementation

The proposed algorithm PPMF, is to be compared with the EE algorithm. The EE algorithm is implemented using C#; the software tool is obtained from the author. The (PPMF) algorithm is, likewise, implemented using C#. The EE algorithm depends on encoding each nucleotide and builds up new data from the original DNA Sequence, using this data to find motifs. The PPMF framework is divided into three stages, with reference to the following merits:

- Minimum hit of pattern per sequence X (default = 2).
- Number of sequences Y = 1.
- Maximum motif length: $K_{(max)} = 4$.
- Motif length: K.
- Pattern P_k.
  - $K = 1 \Rightarrow P_1 = \{A, G, C, T\}$
  - $K = 2 \Rightarrow P_2 = \{AA, AG, AC, AT, GA, GG, GC, GT, \ldots\}$
  - $K = 3 \Rightarrow P_3 = \{AAA, AAG, AAC, AAT, AGA, AGG, AGC, AGT, \ldots\}$
- Pattern $P_K[n] \in P_K \Rightarrow P_2[0] = AA, P_2[1] = AG, P_2[2] = AC, P_2[3] = AT, \ldots$

3.1 The First Stage

1. Scanning the sequence.
2. Finding all occurrences of pattern P where $P \in P_2$.
3. Recording patterns positions in Z with index relative to corresponding pattern index in $P_2 \Rightarrow Z = \{\{\}, \{0, 8, 16, 18\}, \{4, 13, 22\}, \{\}, \{3, 12, 17, 21\}, \{9\}, \ldots\}$ see (Fig. 5(a)).
4. For each pattern P, if the number of occurrences $\geq$ X then print it as 2k motifs element.

3.2 The Second Stage

1. Rescanning the sequence.
2. Finding all occurrences of pattern P where $P \in P_3$.

3. Recording:
   a. patterns positions in Z with index relative to corresponding pattern index in $P_3 \Rightarrow$ Z = {{}, {}, {}, {}, {16}, {8}, {}, {0, 18}, {}, {4}, …} see (Fig. 6).
   b. Next nucleotide in H with index relative to corresponding pattern index in $P_3 \Rightarrow$ H = {{}, {}, {}, {}, {C}, {T}, {}, {C, C}, {}, {T}, …} see (Fig. 6).
4. For each pattern P, if the number of occurrences $\geq$ X then print it as 3k motifs element.

3.3 The Third Stage

The initial input for this stage is the result from the second stage which is 3k motifs, its positions and its next nucleotides.

1. For each motif in P, check the corresponding next nucleotide in H. If it appears at least X time (the minimum number of required occurrence), combine it with the corresponding pattern to get the K + 1 motifs; its positions is available in Z and is known.
2. Rescan the sequence at positions corresponding to K + 1 motifs from the previous step to get the new next nucleotide.
3. Use the K + 1 motifs from previous step, its positions and its next nucleotides as the input data to the third stage. Repeat the steps until getting the required maximum motifs length.

In the EE algorithm, the coding process will be more complex as motif length increases. With the increase in motif length, the same steps on the PPMF algorithm are to be applied. The degree of complexity of the proposed algorithm is, apparently, less than that of the EE algorithm.

4 Experimental Results and Discussion

The proposed PPMF algorithm is applied and compared with the other algorithm on the same pc configurations: Intel core i7 processor running at 2.67 GHz with 8 GB of RAM. The average execution time to find all the identical string motifs of length up to 40 nucleotides of the proposed PPMF algorithm is shown in (Figs. 8, 9, and 10) as compared with the average execution time of the EE algorithm applied on 3 different sizes data sets. The same data used in the EE algorithm acts as the data input of the proposed algorithm. The tool, which runs the PPMF algorithm, is created using C# language [8]. There are three different data sets, extracted from TRANSFAC [9, 10] and can be downloaded [11].

The first data set appears in Fig. 8, the average execution time of the PPMF algorithm decreases by about 68% of the average execution time of the EE algorithm. Worthy of note is that at the nucleotide 3000, a sharp peak appears because the sequence contains only one subsequence, while the next contains more than one.

The second data set appears in (Fig. 9). It is a real biological sequence larger than first data set. The average execution time of the proposed algorithm decreases by about 72.3% of the average execution time of the compared algorithm. As shown in (Fig. 9),

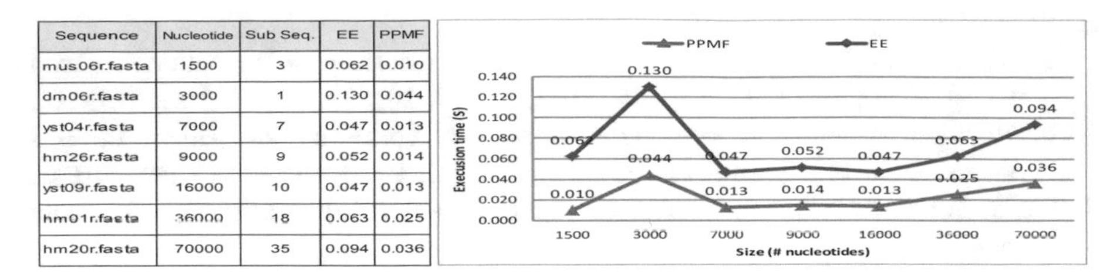

Sequence	Nucleotide	Sub Seq.	EE	PPMF
mus06r.fasta	1500	3	0.062	0.010
dm06r.fasta	3000	1	0.130	0.044
yst04r.fasta	7000	7	0.047	0.013
hm26r.fasta	9000	9	0.052	0.014
yst09r.fasta	16000	10	0.047	0.013
hm01r.fasta	36000	18	0.063	0.025
hm20r.fasta	70000	35	0.094	0.036

Fig. 8. The average execution time (in seconds) to find identical string motifs of length up to 40 nucleotides running on the first data set.

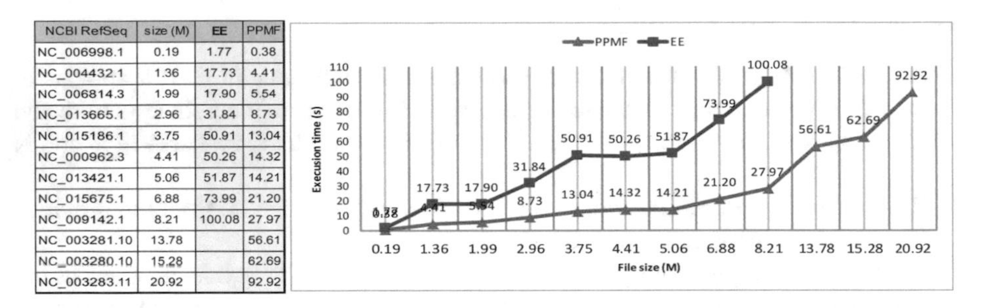

NCBI RefSeq	size (M)	EE	PPMF
NC_006998.1	0.19	1.77	0.38
NC_004432.1	1.36	17.73	4.41
NC_006814.3	1.99	17.90	5.54
NC_013665.1	2.96	31.84	8.73
NC_015186.1	3.75	50.91	13.04
NC_000962.3	4.41	50.26	14.32
NC_013421.1	5.06	51.87	14.21
NC_015675.1	6.88	73.99	21.20
NC_009142.1	8.21	100.08	27.97
NC_003281.10	13.78		56.61
NC_003280.10	15.28		62.69
NC_003283.11	20.92		92.92

Fig. 9. The average execution time (in seconds) to find identical string motifs of length up to 40 nucleotides running on the second data set.

the three data sequences (10, 11 and 12) do not apply to the EE algorithm, and the maximum data size which the EE algorithm supports in our test was the sequence with size 8.21 M.

The third data set appears in (Fig. 10). It is a real biological sequence larger than the second data set. The EE algorithm tool did not work with any of these data sets, albeit the increased memory size up to 16 GB. The EE algorithm tool does not support data with size more than about (10 M) in our test. The proposed algorithm has worked normally with files of size up to 100 M with 8 GB RAM pc, while stopped working with bigger files. Increasing the pc memory to 16 GB, all files are processed.

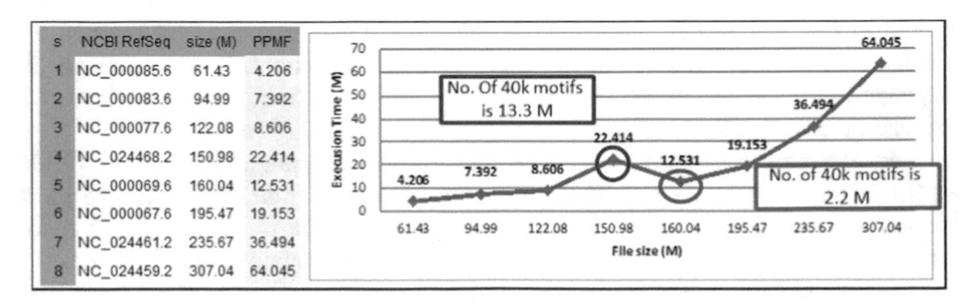

s	NCBI RefSeq	size (M)	PPMF
1	NC_000085.6	61.43	4.206
2	NC_000083.6	94.99	7.392
3	NC_000077.6	122.08	8.606
4	NC_024468.2	150.98	22.414
5	NC_000069.6	160.04	12.531
6	NC_000067.6	195.47	19.153
7	NC_024461.2	235.67	36.494
8	NC_024459.2	307.04	64.045

Fig. 10. The average execution time in minutes to find all identical string motifs of length up to 40 nucleotides in big biological sequences running on third data set.

In (Fig. 10), the average execution time is 64 min to get all identical string motifs up to 40 nucleotides for a sequence with size of 307.04 M of nucleotides, which means that approximately one hour is required to finish processing a sequence of size 300 M. The execution time of the proposed algorithm depends on the size of the sequence data file and the number of motifs in that sequence. The present paper has found out that the data sequence number 4 takes more time than the next sequence, although it is less in size as the number of existing 40k identical string motifs is 13.3 M patterns and is repeated 63.6 M times while in the next sequence, it is 2.2 M motifs and is repeated 14.8 M.

5 Conclusions

In this article, we have proposed an algorithm called (PPMF) for DNA identical string motifs discovery and demonstrated its capacity to discover motifs in real biological datasets. It is designed to support the big data size sequences and is tested with sequence data up to 300 million nucleotides in 60 min of execution time. The execution time depends on the sequence size and number of existing motifs. Comparing, the proposed PPMF algorithm with the EE algorithm, it is discovered that the execution time is less by about 60% to 75% which means that the proposed algorithm is more efficient than the EE algorithm.

The comparative algorithm (EE) encodes the nucleotide to an integer value by using an encoding and decoding schema, the integer values of the nucleotide are stored and then by using sorting and searching algorithms it gets the identical motifs. On the other hand, the proposed algorithm does not depend on the encoding and decoding techniques as on the comparative algorithm. It saves the data in a form, which helps to get the result directly. It assigns each pattern to specified index of an array. It does not need to sort data after reading it. The EE algorithm, saves the each k pattern and its position in a tuple while our proposed algorithm saves the k pattern once and save the corresponding positions, if a pattern of 40 nucleotides is repeated 1000 times, EE algorithm will save these pattern 1000 times, but our algorithm only save the pattern once. This make our algorithm more efficient in reading, writing data, and less memory consuming.

References

1. Rombauts, S., Déhais, P., Van Montagu, M., Rouzé, P.: PlantCARE, a plant cis-acting regulatory element database. Nucleic Acids Res. **27**(1), 295–296 (1999)
2. D'haeseleer, P.: What are DNA sequence motifs? Nat. Biotechnol. **24**, 423–425 (2006)
3. Kaya, M.: Multi-objective genetic algorithm for motif discovery. Expert Syst. Appl. **36**(2), 1039–1047 (2009)
4. Das, M.K., Dai, H.-K.: A survey of DNA motif finding algorithms. BMC Bioinform. **8**(7), S21 (2007). https://doi.org/10.1186/1471-2105-8-S7-S21
5. Karci, A.: Efficient automatic exact motif discovery algorithms for biological sequences. Expert Syst. Appl. **36**(4), 7952–7963 (2009)

6. Al-Ssulami, A., Azmi, A.M.: Towards a more efficient discovery of biologically significant DNA motifs. In: Ortuño, F., Rojas, I. (eds.) Bioinformatics and Biomedical Engineering, pp. 368–378. Springer, Cham (2015)
7. Azmi, A.M., Al-Ssulami, A.: Encoded expansion: an efficient algorithm to discover identical string motifs. PLoS ONE **9**(5), e95148 (2014)
8. Github. https://github.com/abdo5/ppmf.git
9. Wingender, E., Dietze, P., Karas, H., Knüppel, R.: TRANSFAC: a database on transcription factors and their DNA binding sites. Nucleic Acids Res. **24**(1), 238–241 (1996)
10. Tompa, M., Li, N., Bailey, T.L., et al.: Assessing computational tools for the discovery of transcription factor binding sites. Nat. Biotechnol. **23**(1), 137–144 (2005)
11. University of washington, Computer Science & Engineering. http://bio.cs.washington.edu/assessment/download.html

Parallel Density-Based Downsampling of Cytometry Data

Martin Nemček, Tomáš Jarábek(✉), and Mária Lucká

Faculty of Informatics and Information Technologies,
Slovak University of Technology, Bratislava, Slovakia
mrtn.nemcek@gmail.com, {tomas.jarabek,maria.lucka}@stuba.sk

Abstract. Identification of cellular populations is the first step in analyzing cytometry data. To identify both abundant and outlying rare cellular populations a density-based preprocessing of data to equalize representations of populations is needed. Density-based downsampling keeps representative points in the cellular space while discarding irrelevant ones. We propose a fast and fully deterministic algorithm for density calculation, based on space partitioning, tree representation and an iterative approach to downsampling utilizing fast calculation of density. We compared our algorithm with SPADE, the most used approach in this area, achieving comparable results in a significantly shorter runtime.

Keywords: Downsampling · Cytometry · Density ·
Parallel computing · Outlier detection

1 Introduction

Cytometry focuses on measurement and analysis of multiple parameters of cells. The ability to analyze cells has various applications in many fields, identifying and quantifying immune cell populations for monitoring of a patient's immune system and novel biomarker detection [8]. For the cell parameter measurement, mass cytometry which measures up to 40 parameters per cell is used [9,10].

A traditional approach to the analysis of cytometry data is manual gating – identification of the cellular populations from two–dimensional dot plots. This method has drawbacks such as subjectivity, irreproducibility of results and a high time consumption thus approaches automating the process were developed [5, 11, 12]. Generally the results of the analysis are used for identification of dense regions of usual cell populations and outlying rare cell populations.

Multiple software tools such as PhenoGraph [4], viSNE [1] or Citrus [2] tackle this issue differently, however all have their disadvantages, such as losing some information in the data, being ineffective on big data sets or introducing high stochasticity. SPADE [7] is one of the most used software tools for analyzing cytometry data. It uses hierarchical clustering to mirror the structure of cells,

F. Fdez-Riverola et al. (Eds.): PACBB 2019, AISC 1005, pp. 87–95, 2020.
https://doi.org/10.1007/978-3-030-23873-5_11

however a density-based downsampling for preprocessing is performed, to equalize representations of abundant and rare cell populations. To address the stochasticity of results a new version of SPADE was implemented, with a deterministic approach to density-based downsampling that is ineffective on big data sets [6].

A measurement of cell parameters can be viewed as a point in a point cloud. Density-based downsampling is performed based on densities of points. Density of a point is equal to the number of points in its ε–neighbourhood. Point cloud areas with high and low densities correspond to abundant and rare cell populations respectively [7]. After the calculation of densities, density-based downsampling can be performed so that only representative points are kept. The idea is to choose the most representative point, make a "hole" of the size ε with the center at the chosen point and repeating this until the desired result is reached [13].

The time complexity of such an approach depends on the time complexity of density calculation. Some methods [7] utilize stochasticity to tackle the problem of time complexity, render results irreproducible [6]. A deterministic version exists but suffers from high time complexity [6]. To improve both density calculation and downsampling, we propose a deterministic algorithm producing consistent results across multiple runs while enabling the usage on big datasets.

The paper is organized as follows. Section 2 describes the proposed algorithm for density calculation and the iterative density based downsampling as well as the parallelization of the algorithm. Section 3 presents experimental results and describes used datasets. The paper concludes with Sect. 4.

2 Proposed Algorithm

We propose a fully deterministic fast algorithm for the calculation of density using a tree representation of the partitioned space and an iterative approach to the density-based downsampling utilizing a fast calculation of density.

2.1 Density Calculation

The density of a point is the number of points in its ε–neighborhood. The size of the neighborhood is denoted by parameter ε. To calculate the density of a point we need to know its distance to all other points to determine whether they are neighbors. This approach is computationally demanding on big datasets. The main idea of our algorithm is the minimization of the number of points used to find the densities. The L_1 distance metric is used to compute the distance as used in SPADE [7].

Partitioning the space in each dimension to intervals of size ε, reduces the number of possible neighbor points. If we consider a two dimensional space (Fig. 1) normalized to $[0, 1]$ and partitioned to intervals we get a grid of squares of size $\varepsilon \times \varepsilon$. For any point p in square c, neighbor points of point p are only in the neighbor squares of c. Thus, to calculate the density of point p we need to calculate distances only to points of the neighbor squares of p.

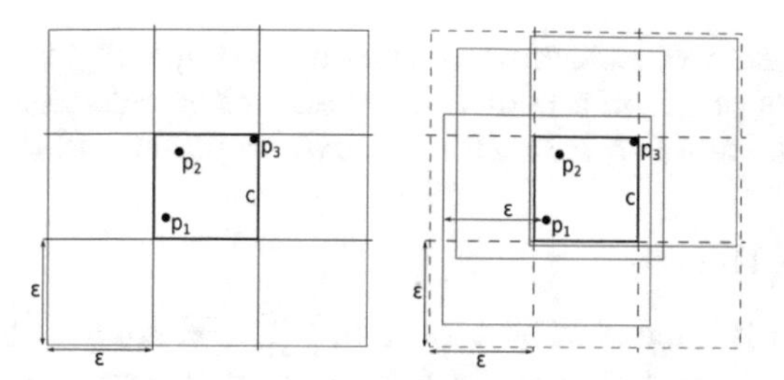

Fig. 1. On the left: 2D space partitioned to intervals of size ε. On the right: The same partitioned space with ε- neighborhoods of the shown points.

We denote the m-dimensional hypercube as a square. Points are placed into the squares based on their coordinates at each dimension. An index of ε-interval is calculated at each dimension to get the position of the square by formula

$$\varepsilon(p, d) = \left\lfloor \frac{p_d}{\varepsilon} \right\rfloor$$

where p is a point in space, d is an index of the dimension for which the ε-interval is calculated and p_d is the coordinate of point p at dimension d.

If ε splits each of m dimensions to k intervals it creates k^m squares and with a rather small m we get $n \ll k^m$ where n is number of points. Most squares would be empty while having enormous memory requirements. We propose a tree representation called a density tree to represent such a space and utilize space partitioning by ε to calculate the point densities. Each level defines one dimension and leaf nodes represent squares of the partitioned space. Every node contains a map of the ε-interval indices of the next dimension. The density tree is built sequentially as new points are pushed into it. When a new point is pushed into the tree we iterate over the dimensions of the point and for each dimension we calculate the ε-interval index and move the point from the node representing the ε-interval in the current dimension to the next node. If such a node does not exist then it is created. This sequential building of the density tree ensures that only non-empty cells are represented. To illustrate this, Fig. 2 presents a partitioned space and the density tree built from the partitioned space. As can be seen on Fig. 2 only subspaces containing at least one point are represented as leaf nodes in the density tree.

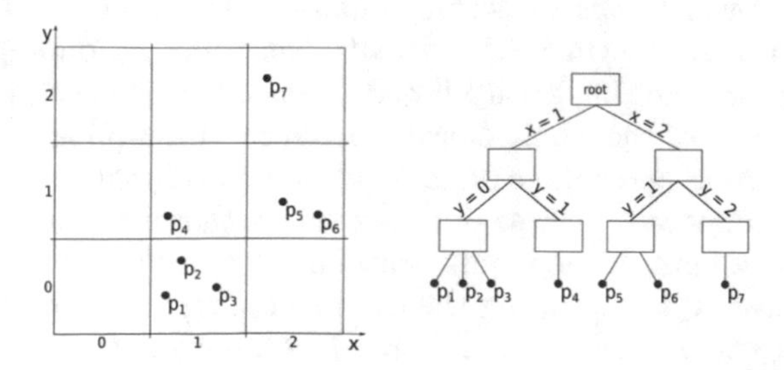

Fig. 2. Example of density tree built from a partitioned space with four leaf nodes.

By parallelization we can improve the time complexity because the calculation of densities of points of a node is independent on calculation of other points and nodes and therefore so for each node can be performed simultaneously.

2.2 Weighted Density

Density of a point is calculated by a formula $\sum_1^q 1$, where q is number of neighbor points. Using such calculation of density can lead to some misinterpretations.

Therefore we propose a weighted density where the contribution of a neighbor point to the density of a point is based on their distance and is calculated as $c = (1 - \frac{d}{\varepsilon})^e$, where c is the contribution, d is the distance between the neighbor points, ε is the size of the neighborhood and e is the Euler's number. Accordingly, the weighted density w of a point p, is calculated as

$$w(p) = \sum_{i=1}^{M} \left(1 - \frac{d(p, q_i)}{\varepsilon}\right)^e$$

where M is number of the neighbor points of point p and $d(p, q_i)$ is the distance between point p and its i-th neighbor point q_i.

The main purpose of using a weighed density is to ensure that closer neighbor points will contribute more to the final weighted density of the point than neighbor points which lie near the edge of the neighborhood. We chose the aforementioned formula due to its properties which reflect the desired relationships between point and inside its neighbourhood. Choosing different formula or other value for e can put more emphasis on weight contribution by neighbour points.

2.3 Iterative Density-Based Downsampling

The representativeness of a point in our algorithm is equal to the weighted density of that point (see Sect. 2.2) meaning that a point is more representative the more closer neighbor points it has. The process of downsampling aims to find the most representative point among the points which were not chosen to be kept or discarded yet and keeping that point and discarding all its neighbor points and iterating until all the points are either kept or discarded.

The time complexity of the density-based downsampling is largely affected by the time complexity of the density calculation because the downsampling is fast in comparison to the density calculation. Choosing the incorrect value of ε can lead to a slowdown in the density-based downsampling. We propose an iterative approach to the density-based downsampling which vastly improves the time complexity of the whole density-based downsampling process and only requires the resulting percentage of kept points as an input parameter.

The iterative approach utilizes the observation that if a relatively small value of ε is chosen then most of the points would end up in their own leaf nodes with exception to dense regions. During the downsampling only the irrelevant points from the dense regions would be discarded because most of the points would have few to non neighbor points.

The aforementioned observation is utilized in such a way that a relatively small value of ε is chosen, the densities are then calculated and the points are density-based downsampled. The algorithm terminates if $c = r \pm \delta$ where c is current percentage of the kept points, r is selected resulting percentage and δ is a chosen constant which adjusts the speed and precision of the algorithm. If value of δ is too big, algorithm needs less iterations to terminate but the result might not be as precise as desired. On the other hand if value of δ is too small e.g. almost 0 then algorithm needs more iterations to meet the terminating condition. We remark that the value of ε iteratively increases according to the following expression. Value of ε is adjusted as:

$$\varepsilon = \begin{cases} \varepsilon \times 2, & c > r \\ \varepsilon_p + \frac{|\varepsilon - \varepsilon_p|}{2}, & otherwise \end{cases}$$

where ε_p is value of ε from the previous iteration when $c > r$. By increasing the value of ε in the next iteration more points are discarded and fewer are kept resulting in descending to the resulting percentage otherwise by decreasing the value of ε fewer points are discarded and more are kept resulting in ascending to resulting percentage. Also if $c > r$ then only the kept points from the current iteration are used as input data to the next iteration otherwise if $c \leq r$ then the points from last iteration where $c > r$ was true are used as input data to the next iteration. Then the algorithm continues with calculating the densities and downsampling on the new data from the previous iteration.

To improve results we perform weight adjustment. To the value of the weighted density of each point, half of the weighted density of that point from the previous iteration is added. This step helps to minimize the number of "holes" in the resulting space containing no points while there should be at least one point.

2.4 Parallelization of the Algorithm

The density calculations of points of a density tree's leaf node is an independent process and therefore can be computed in parallel. We are able to compute simultaneously densities of NP nodes where NP is the number of virtual processors, whereby we consider the shared memory model of computing.

To improve the performance of computing, the distance between two points is calculated only once at each node. To address the issue of race conditions we identified the information about computed distances between nodes. Each thread decides either to terminate or to continue as a critical section. Hence, only one thread manipulates that information and multiple calculations of distances between two points are prevented.

The time complexity depends on the number of points at each node and therefore if one node contains many points the time complexity of the entire process is heavily affected. The calculation of the point densities is a totally independent process so they can also be computed in parallel.

3 Experimental Results

We compared our proposed algorithm with SPADE [6] which is the most used software tool for the density-based downsampling of cytometry data. We compared the time needed for density calculation and density-based downsampling. We used several datasets with a varying number of points including smaller and bigger datasets, with 13 dimensions.

We used two datasets (i) dataset from The Slovak Academy of Sciences and (ii) a mouse bone marrow sample with known cellular hierarchy [7]. The dataset available from The Slovak Academy of Sciences contained mass cytometry data with 45 markers from 855 samples coming from patients having either multiple myeloma, Waldenström myeloma or were healthy. Dataset samples varied greatly in size ranging from 1 667 to 139 246 956 events. Individual samples were categorized into three categories (panels) - P2, P3 and P4. Our experimental results were performed on samples from panel P2 and P3 using 13 shared markers.

Table 1. Time in minutes needed for the calculation of densities for the dataset from The Slovak Academy of Sciences.

n	~81 k	~248 k	~382 k	~491k	~624 k
SPADE	0.683	7.65	16.883	26.36	38
densamp	0.666	3.35	3.116	21.26	18

Table 2. Time in minutes needed for the density-based downsampling for dataset from The Slovak Academy of Sciences.

n	~81 k	~248 k	~382 k	~491 k	~624 k
SPADE	0.939	9.732	21.783	38.95	58.083
densamp	0.156	0.689	0.84	3.419	2.024

To confirm the biological correctness, we present results of the analysis on the mouse bone marrow samples processed by our algorithm and SPADE.

We compared the performance of the density calculation of both SPADE and our algorithm *densamp*. To get consistent results, we let SPADE calculate the value of ε which was used for density calculation by SPADE and *densamp*. Table 1 presents computational times in minutes needed for calculations of densities by SPADE and *densamp* using the same value of ε and various number of points. For smaller datasets, differences are not significant, but they become significant with bigger datasets. We can see that the runtime is not straightforwardly dependent on the size of data, because it also depends on the topology of the input space (see Sect. 2.1). Our algorithm is overall faster than SPADE for all tested datasets.

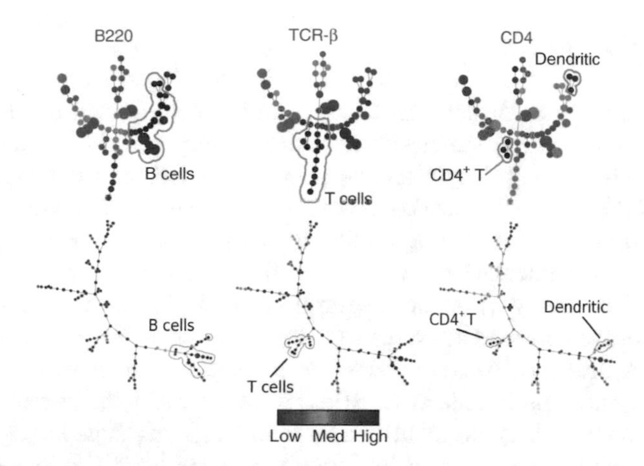

Fig. 3. Identification of dendritic cell population in the mouse bone marrow sample by SPADE and our algorithm. Upper three visualizations were produced by SPADE and the lower three were produced by our algorithm.

We also compared the times needed to perform the density-based downsampling. We chose $r = 10\%$ as the resulting percentage of points after downsampling for both SPADE and *densamp*. Table 2 shows the time in minutes needed to perform density-based downsampling. We can see the significant differences between SPADE and our algorithm. The differences are caused by the iterative approach which calculates densities with a relatively small value of ε at first. The value of ε iteratively increases while reducing the amount of processed data, further reducing the time for the calculation of densities while approaching to r. Overall runtime of our algorithm is up to 30 times faster than SPADE for large datasets.

The acquired speed up of the process of density based downsampling would be irrelevant if the results were incorrect. We compared results on the mouse bone marrow sample with a known cellular hierarchy. The gained biological results were confirmed by domain experts from the Slovak Academy of Sciences.

Authors of SPADE identified a cell population of dendritic cells using SPADE which was not identified by manual gating. Dendritic cell population is defined as TCR-β^- B220$^+$ CD4$^+$ [7]. We processed the cytometry data using the same 8 cytometry markers as authors of SPADE. An *arcsinh* transformation with value 150 of the cofactor parameter was applied before performing the density-based downsampling to the desired 10%.

Downsampled data was clustered into 100 clusters, upsampled and visualized in a tree structure (Fig. 3). For visualization a non-deterministic Fruchterman-Reingold algorithm [3] was used hence results by *densamp* and SPADE differ topologically, but visualize the same information. Our algorithm correctly identified the cell population of dendritic cells contrary to manual gating.

Weight adjustment during the density-based downsampling (see Sect. 2.3) is crucial to correctly identify rare cell populations, else rare cell populations could be discarded during downsampling and hence not included in the results.

4 Conclusions

Existing approaches to density calculation and density-based downsampling of cytometry data use either stochastic methods to improve their time complexity or perform badly on big data sets. We proposed a fast and deterministic algorithm for calculations of densities based on a space partitioning using a tree representation and an iterative approach to density-based downsampling utilizing the fast density calculation. We compared our algorithm with the most used software tool for density-based downsampling SPADE. Our algorithm had a significantly improved needed time also on big data sets. We also compared our results from a biological point of view on a mouse bone marrow sample with known cell populations. Similarly as SPADE we were able to identify the outlier cell population which was not identified by manual gating. The biological correctness of the results was confirmed by domain experts from the Slovak Academy of Sciences. In the iterative approach to the density based downsampling the value change of the ε parameter is a geometric series. This results in increased time complexity of some iterations. We understand that the choice of using the density tree can be replaced by using a hash resulting in similar results but have chosen the density tree to represent the data because tree data structures are a well researched problem.

Acknowledgments. This work was partially supported by the Scientific Grant Agency of The Slovak Republic, Grant No. VG 1/0458/18 and APVV-16-0484.

References

1. Amir, E.A.D., et al.: viSNE enables visualization of high dimensional single-cell data and reveals phenotypic heterogeneity of leukemia. Nat. Biotechnol. **31**(6), 545–552 (2013)
2. Bruggner, R.V., et al.: Automated identification of stratifying signatures in cellular subpopulations. Proc. Natl. Acad. Sci. **111**(26), E2770–E2777 (2014)
3. Fruchterman, T.M.J., Reingold, E.M.: Graph drawing by force-directed placement. Softw. Pract. Exper. **21**(11), 1129–1164 (1991)
4. Levine, J., et al.: Data-driven phenotypic dissection of aml reveals progenitor-like cells that correlate with prognosis. Cell **162**(1), 184–197 (2015)
5. Li, H., et al.: Gating mass cytometry data by deep learning. Bioinformatics **33**(21), 3423–3430 (2017)
6. Qiu, P.: Toward deterministic and semiautomated SPADE analysis. Cytometry. Part: J. Int. Soc. Anal. Cytol. **91**, 281–289 (2017)
7. Qiu, P., et al.: Extracting a cellular hierarchy from high-dimensional cytometry data with SPADE. Nat. Biotechnol. **29**(10), 886–891 (2011)
8. Saeys, Y., et al.: Computational flow cytometry: helping to make sense of high-dimensional immunology data. Nat. Rev. Immunol. **16**(7), 449–462 (2016)
9. Spitzer, M., Nolan, G.: Mass cytometry: single cells, many features. Cell **165**(4), 780–791 (2016)
10. Tanner, S.D., et al.: An introduction to mass cytometry: fundamentals and applications. Cancer Immunol. Immunother. **62**(5), 955–965 (2013)

11. Verschoor, C.P., et al.: An introduction to automated flow cytometry gating tools and their implementation. Front. Immunol. **6**, 380 (2015)
12. Weber, L.M., Robinson, M.D.: Comparison of clustering methods for high-dimensional single-cell flow and mass cytometry data. Cytom. Part A **89**(12), 1084–1096 (2016)
13. Zare, H., et al.: Data reduction for spectral clustering to analyze high throughput flow cytometry data. BMC Bioinform. **11**(1), 403 (2010)

Signaling Transduction Networks in Choroidal Melanoma: A Symbolic Model Approach

Beatriz Santos-Buitrago[1(✉)] and Emiliano Hernández-Galilea[2,3]

[1] Bio and Health Informatics Lab, Seoul National University, Seoul, South Korea
bsantosb@snu.ac.kr

[2] Department of Ophthalmology, University Hospital of Salamanca, Salamanca, Spain
egalilea@usal.es

[3] Institute for Biomedical Research Salamanca (IBSAL), Salamanca, Spain

Abstract. Biochemical reactions that take place concurrently in a cell can be explored and analyzed by symbolic systems biology. These cellular processes can be modeled with symbolic mathematical models through the use of rewrite rules. Our goal is to define formal models that capture biologists intuitions and reasoning. Pathway Logic is a system for developing executable formal models of biomolecular processes. Analyses of biological facts can be obtained from such models. Ocular melanoma is the most frequent malignant primary intraocular tumor in adult population and the second most common site of malignant melanoma in the body. The knowledge of the signaling pathways involved in melanoma offers new treatment strategies. In this paper, we provide a symbolic system that explores complex and dynamic cellular signaling processes that induce cellular proliferation and survival in choroidal melanoma.

Keywords: Symbolic systems biology · Choroidal melanoma · Signal transduction · Rewriting logic · Pathway Logic

1 Signaling Pathways in Choroidal Melanoma

Complex biological mechanisms can be deciphered in an in-depth manner through their structure, dynamics, and control methods. Predictive models would greatly benefit the research of human signaling processes. These molecular pathways detect cells, transform components, and internally transmit information from their environment to intracellular targets, such as the genome [22].

There are a large number of qualitative approaches for computational analysis of cellular signaling networks, such as ordinary differential equations [16,18]. Quantitative analyses deal with large numbers of molecules per species, but their complexity grows enormously when this number is huge. Qualitative modeling provides alternative approaches when qualitative methods are not possible.

F. Fdez-Riverola et al. (Eds.): PACBB 2019, AISC 1005, pp. 96–104, 2020.
https://doi.org/10.1007/978-3-030-23873-5_12

Symbolic models allow us to model, compute with, analyze, and reason about networks of biomolecular interactions at multiple levels of detail, depending on the information available and the issues to be researched. Such models may suggest new insights and understandings of complex biological processes. This formalism provides a language for representing system states and change mechanisms (such as reactions), and analysis tools based on computational or logical inference. Symbolic models can simulate the behavior of a system. The goal is to define formal models closer to the biologists mindsets [21]. Rule-based models can manage biological interactions in a natural manner. These cellular processes are also treated satisfactorily and efficiently thanks to the ability of the rule-based systems to deal with systems of great subjacent complexity [5].

In this work, we want to define rule-based systems to analyze the evolution of different initial states and to study the reachable states from these initial states in main signaling pathways involved in choroidal melanoma.

Ocular melanoma develops from the melanocytes of the uvea and reaches 5% of the total melanomas, being the most frequent malignant primary intraocular tumor in adult population [1]. Uveal melanoma is an aggressive neoplasm with an overall mortality at approximately 15 years of 50% whatever treatment is applied [3,4]. The incidence of uveal melanoma varies between 4.3 and 10.9 cases per million inhabitants per year.

These tumors can be discovered in a routine exploration without symptoms or it can be presents with symptoms as a loss of visual acuity or visual field. Diagnostic accuracy has been shown to be greater through ophthalmoscopy and ocular ultrasound. The clinical and histological properties of uveal melanoma have been well identified and the most aggressive tumor types have been established. Progress has also been made in molecular pathogenesis and cytogenetics to predict the risk of metastasis [7,9,12].

Microenvironment are present in this tumor. Different cells make up this microenvironment and communicate each with other through direct or cytokines and chemokines production and acting in an autocrine or paracrine manner. Cytogenetic analysis in uveal melanomas has shown chromosomal changes that give rise to regulatory oncogenes or tumor suppressors. Cells communicate with each other and respond to environmental stimuli through transduction pathways. Alteration of these pathways leads to anomalous cellular responses as occurs in cancer. In recent years, there has been progress in the knowledge of the molecular biology of cancer and in the pathological implications of some integrated molecules in the signaling pathways [19].

The rest of the paper describes the main characteristics of underlying rewriting logic and Pathway Logic in Sect. 2. Some notes of the modeling of signaling networks with rewriting logic are presented in Sect. 3. The analysis of the dynamics in our melanoma case study is developed in Sect. 4. We draw our conclusions in Sect. 5.

2 Rewriting Logic and Pathway Logic

Rewriting logic constitutes a logic of change or becoming [11]. It facilitates the specification of the dynamic features of systems and naturally deals with highly nondeterministic concurrent computations. Rewriting logic has good properties as a flexible and general semantic framework for giving semantics to a broad spectrum of languages and models of concurrency [17]. Rewriting logic is efficiently implemented in Maude language [2].

An equational theory defines a rewriting logic theory, so that you can define sorts, constructors, function symbols, and equality between terms. Rewrite rules of rewriting logic extends equational theory, so that the dynamics between states can be represented. Rewrite rules establish local and parallel changes in a dynamic concurrent system. In this way, these deduction rules allow sound reasoning. In a logical language, each rewrite rule is a logical entailment in a formal system.

Pathway Logic [20] is an approach to the modeling and analysis of molecular and cellular processes based on rewriting logic. Using the Maude system, the resulting formal models can be executed and analyzed [15]. The naturalness of rewriting logic for modeling and experimenting with mathematical and biological problems has been illustrated in a number of works [10]. Pathway Logic system has been used to curate models of signal transduction [14].

A *rule knowledge base* consists of several rewrite rules together with the supporting data type specifications. A *model* of melanoma signaling system consists of a specification of an *initial state* (cell components with their locations) and a collection of rules derived from the global knowledge base by a symbolic reasoning process that looks for all rules that could possibly be applicable in an execution starting from the initial state. Such executable models reflect the possible ways a system can evolve. Logical inference of Pathway Logic can: (1) simulate possible ways a system could evolve, (2) build pathways in response to queries, and (3) think logically about dynamic assembly of complexes and cascading transmission of signals.

Data types like proteins and genes are defined as Maude `sorts`, while functions on these sorts are defined by means of `op`. For example, we define constants for mitogen-activated protein kinase 3 (`Erk1`) and RAC-α serine/threonine-protein kinase (`Akt1`) as:

```
sorts Protein Gene .   ops Erk1 Akt1 : -> Protein [ctor] .
```

where the attribute `ctor` indicates that this constant is *constructor*. Relations between sorts are stated by means of subsorts.

3 Case Study I: Modeling of Signaling Pathways in Choroidal Melanoma

Different elements that appear in a cell (proteins, genes, and so on) are defined as a `Soup` (i.e. an associative, commutative list with a neutral element). These elements form a location, which is identified by a location name (`LocName`):

```
op {_|_} : LocName Soup -> Location [ctor] .
```

Different elements appear in different parts or locations of the cell: outside the cell (XOut), in/across the cell membrane (CLm), attached to the inside of the cell membrane (CLi), in the cytoplasm (CLc), and in the nucleus (NUc). We can indicate that nucleus NUc contains proteins and genes Tp53-gene (gene transcription is on), Rb1, Myc, and so on as:

```
{NUc | [Tp53-gene - on] Rb1 Myc Maz Chek1 Chek2 Tp53 NProteasome}
```

Finally, dishes are defined as wrappers of Soup, which in this case are not isolated elements but different locations:

```
op PD : Soup -> Dish [ctor] .
```

A cell would be represented in Maude with the following *tiny* dish:

```
op TinyDish : -> Dish .
eq TinyDish = PD( {XOut | Igf1 } {CLm | Igf1R Erbb2 [Cbl - Yphos]}
 {CLi | [Hras - GDP] [Gnai1 - act]} {NUc | Maz Msk1 Elk1}
 {CLc | [Gsk3s - act] [Csnk1a1 - act] Akts Mek1} ) .
```

Figure 1 shows a schematic representation of this cell. Some proteins are represented: insulin-like growth factor I (Igf1), receptor tyrosine-protein kinase erbB-2 (Erbb2), Myc-associated zinc finger protein (Maz), etc. Some components appear with different modifiers: activation (act), phosphorylation on tyrosine (Yphos), or binding to GDP (GDP). A ligand/receptor binding between elements can be defined with operator (_:_). For example, a binding between Egf and EgfR is written as (Egf : EgfR).

We define a dish SKMEL133Dish for our case study that contains cell locations (such as CLm) and each location contains their elements (such as PIP2) with their corresponding modification states (e.g. [Rheb - GTP], [Ilk - act], etc.). A simplified version of the dish, i.e. some of the elements in the cytoplasm that are not significant are indicated with some ellipses, is:

```
op SKMEL133Dish : -> Dish .
eq SKMEL133Dish = PD( {XOut | empty} {CLo | empty} {CLm | PIP2}
 {CLi | Parva Pi3k Pld1} {CVc | (Tsc1 : Tsc2) [Rheb - GTP]}
 {CLc | [Csnk1a1 - act] [Gsk3s - act] [Ilk - act] Akts BrafV600E Erks Irs1
  Mek1 Mtor Pdpk1 Pkca Rsk1 Akt1 Cdc42 Erbb2 Igf1R NgfR ...} {Sig | empty}
 {NUc | Rb1 Chek1 Chek2 Myc Tp53 NProteasome Maz [Tp53-gene - on]} ) .
```

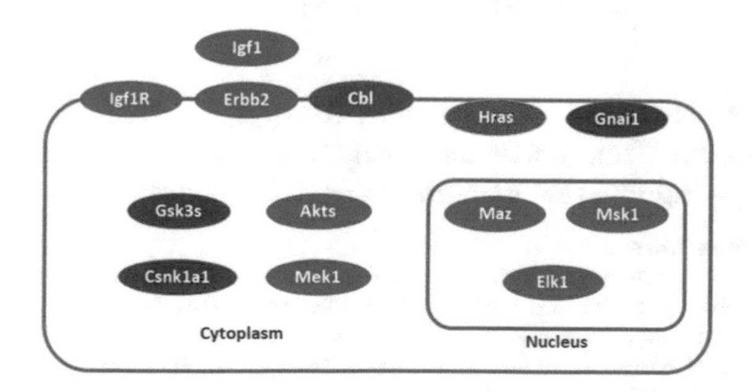

Fig. 1. Schematic representation of a cell. Proteins without modifications are shown in green. The activated proteins are marked in red, those phosphorylated in blue, and those bound to GDP in yellow.

4 Case Study II: Rule-Based Dynamics in Choroidal Melanoma

The *MAPK*, *PI3K*, *mTOR*, and *IGF-1R* pathways have been actively involved in the choroidal melanoma [8,13]. Initially, a G protein-coupled receptor can activate an associated G protein by exchanging the GDP bound to the G protein for a GTP on the Gα subunit. Gα-GTP mediates activation of PLCβ and promotes cleavage of PIP2 into the second messengers IP3 and DAG. Then, DAG diffuses along the plasma membrane and can activate any protein kinase C (`PKC`). In this way, MAPK signaling would lead to a situation of tumor growth and proliferation [13].

On the other hand, the GNAQ Q209L mutation inactivates the phosphatase of the Gα protein and triggers the MAPK signaling. PI3K also intervenes in phosphorylation of PIP2 to PIP3, and PTEN is opposed to this process. PIP3 then activates AKT and the mTOR-signaling pathway, which leads to cellular growth and proliferation [6].

The IGF-1R signaling is achieved by the inhibition that mTOR produces on the IGF-1 receptor. Moreover, IGF-1 binds to the IGF-1 receptor and leads to IRS activation, which can then activate both the PI3K and MAPK pathways. These signaling pathways can also be altered in other places (e.g. rapamycin, AZD6244, 17-DMAG, and so on).

The rewriting rules allow us to define the concurrent cellular reactions. For example, rule `3820c` establishes that `Pi3k` the inside of the cell membrane mediates phosphorylation of `PIP2` to `PIP3` in the cell membrane (the variables `clm` and `cli` can be replaced by any element soup; see Fig. 2):

```
rl[3820c.PIP3.from.PIP2]: {CLm | clm PIP2 }  {CLi | cli Pi3k }
=> {CLm | clm PIP3 }  {CLi | cli Pi3k } .
```

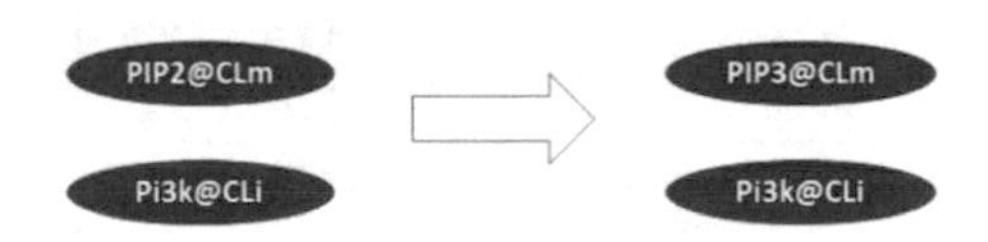

Fig. 2. Schematic representation of the rule `3820c.PIP3.from.PIP2`.

Now, we use the `rewrite` command to apply rewrite rules and obtain a reachable state from our particular dish. The following example provide the result after applying five rewrite steps to our initial dish:

```
Maude> rewrite [5] SKMEL133Dish .
result Dish: PD( {Sig | empty}  {CLm | PIP3}  {CLi | Parva Pi3k Pld1}  {CLo | empty}  {XOut | empty}
{CVc | (Tsc1 : Tsc2) [Rheb - GTP]}  {NUc | Chek1 Chek2 Maz Myc NProteasome Rb1 Tp53 [Tp53-gene - on]}
{CLc | Akts Akt1 Ang Axin1 Baiap2l1 Bim BrafV600E Btrc C10orf90 Cbp Cdc42 Cdk5rap3 Cdkn2a Ctnnb1 Cul5
Cul7 Dzip3 Eif4ebp1 Ep300 Erbb2 Erks Erk5 Fbxw8 G3bp1 Gli1 Huwe1 Igf1R Irs1 Magea3 Mdm2 Mdm4 Mek1
Mkrn1 Mlst8 Mtor NgfR Nr0b2 Nus1 Pax3 Pcmt1 Pdlim7 Pkca Pkch Ppm1d Proteasome Psme3 Rac1 Rad54b
Raptor Rbbp6 Rbx1 Rchy1 Rictor Rnf6 Rnf31 Rnf43 Rps6 Rsk1 S6k1 Sin1 Skp1 Stub1 Syvn1 Tax1bp3 Tpt1
Trim24 Trim28 Ubc13 Ube2d1 Ube2d2 Ube2d3 Ube3a Ube4b Ybx1 Ywhas [Csnk1a1 - act] [Gsk3s - act]
[Ilk - act] [Pdpk1 - act]})
```

Some proteins (e.g. Csnk1a1, Gsk3s, Ilk, and Pdpk1) are activated in the cytoplasm for this possible first solution. However, several rules can be applied to the same dish and these rules will obtain different results.

Figure 3 shows the main signaling pathways that have influence in choroidal melanoma. To obtain more information, search command performs a breadth-first search looking for the pattern given in the command. For example, we can find out if a dish with protein Pi3k stuck to the inside of the plasma membrane and with protein Erks activated in the cytoplasm or nucleus is reachable from SKMEL133Dish in eight steps:

```
search [2,8] SKMEL133Dish  =>*  PD( {loc:LocName | things:Things [Erks - erksmodset:ModSet act]}
   {NUc | nuc:Things } {CLm | clm:Things PIP3 }  {CLi | cli:Things Pi3k }  S:Soup )
   such that  (loc:LocName == CLc)  or  (loc:LocName == NUc) .
```

where the variable S:Soup abstracts the rest of elements, which are not relevant in this case, and the search option =>* stands for zero or more steps. Maude provides two possible solutions which fulfill this condition and it also shows the terms that match in the solution.

```
Solution 1 (state 186)
S:Soup -->  {Sig | empty}  {CLo | empty}  {XOut | empty} {CVc | (Tsc1 : Tsc2) [Rheb - GTP]}
clm:Things --> empty   cli:Things --> Parva Pld1
nuc:Things --> Chek1 Chek2 Maz Myc NProteasome Rb1 Tp53 [Tp53-gene - on]
loc:LocName --> CLc   erksmodset:ModSet --> phos(TEY)
things:Things --> Akts Akt1 Ang Axin1 Baiap2l1 Bim Btrc C10orf90 Cbp Cdc42 Cdk5rap3 Cdkn2a
Ctnnb1 Cul5 Cul7 Dzip3 Eif4ebp1 Ep300 Erbb2 Erk5 Fbxw8 G3bp1 Gli1 Huwe1 Igf1R Irs1 Magea3
Mdm2 Mdm4 Mkrn1 Mlst8 Mtor NgfR Nr0b2 Nus1 Pax3 Pcmt1 Pdpk1 Pdlim7 Pkca Pkch Ppm1d Proteasome
Psme3 Rac1 Rad54b Raptor Rbbp6 Rbx1 Rchy1 Rictor Rnf6 Rnf31 Rnf43 Rps6 Rsk1 S6k1 Sin1 Skp1
Stub1 Syvn1 Tax1bp3 Tpt1 Trim24 Trim28 Ubc13 Ube2d1 Ube2d2 Ube2d3 Ube3a Ube4b Ybx1 Ywhas
[Braf - act] [Csnk1a1 - act] [Gsk3s - act] [Ilk - act] [Mek1 - act phos(SMANS)]
```

In addition, these are the rules that have been applied to reach state 186:

```
Maude> show path labels 186 .
3820c.PIP3.from.PIP2 3808c.BrafV600E.act 431c.Mek1.by.Braf 014c.ErkS.by.Mek1
```

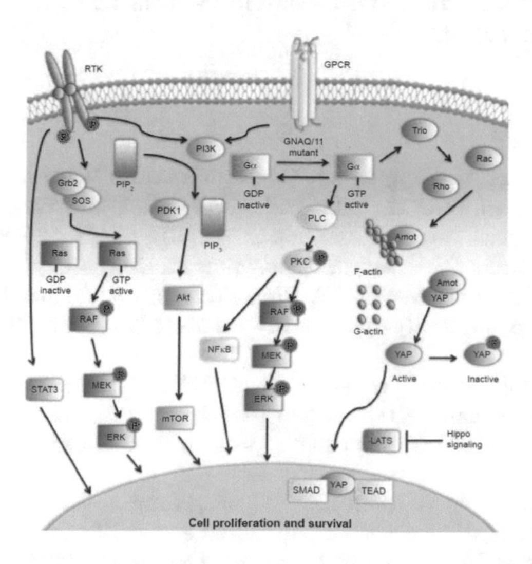

Fig. 3. Signaling pathways in uveal melanoma. (adapted from [6])

The second solution that has been found is:

```
Solution 2 (state 457)
S:Soup --> {Sig | empty} {CLo | empty} {XOut | empty} {CVc | (Tsc1 : Tsc2) [Rheb - GTP]}
clm:Things --> empty   cli:Things --> Parva Pld1
nuc:Things --> Chek1 Chek2 Maz Myc NProteasome Rb1 Tp53 [Tp53-gene - on]
loc:LocName --> CLc   erksmodset:ModSet --> phos(TEY)
things:Things --> Akts Akt1 Ang Axin1 Baiap2l1 Bim Btrc C10orf90 Cbp Cdc42 Cdk5rap3 Cdkn2a
Ctnnb1 Cul5 Cul7 Dzip3 Eif4ebp1 Ep300 Erbb2 Erk5 Fbxw8 G3bp1 Gli1 Huwe1 Igf1R Irs1 Magea3
Mdm2 Mdm4 Mkrn1 Mlst8 Mtor NgfR Nr0b2 Nus1 Pax3 Pcmt1 Pdlim7 Pkca Pkch Ppm1d Proteasome
Psme3 Rac1 Rad54b Raptor Rbbp6 Rbx1 Rchy1 Rictor Rnf6 Rnf31 Rnf43 Rps6 Rsk1 S6k1 Sin1 Skp1
Stub1 Syvn1 Tax1bp3 Tpt1 Trim24 Trim28 Ubc13 Ube2d1 Ube2d2 Ube2d3 Ube3a Ube4b Ybx1 Ywhas
[Braf - act] [Csnk1a1 - act] [Gsk3s - act] [Ilk - act] [Mek1 - act phos(SMANS)] [Pdpk1 - act]
```

In this new solution, it is verified that protein `Erks` is also phosphorylated in `TEY` domain in addition to activated (see values in `erksmodset:ModSet`).

5 Conclusions

Complex biological mechanisms can be deciphered in an in-depth manner through their structure, dynamics, and control methods. Predictive models would greatly benefit the research of human signaling processes. These molecular pathways detect cells, transform components, and internally transmit information from their environment to intracellular targets, such as the genome [22]. Qualitative approaches for computational analysis has shaken up the research in cell biology and medical biology. Formalism of symbolic models consists of: (1) a language to represent the system states; (2) mechanisms to model their alterations; and (3) analysis tools based on logical inference [18].

In this paper, we present a computational analysis of signaling pathways based on rewriting logic. Protein activation of the families `ERK` and/or `AKT` characterizes cellular proliferation and survival [6,14]. We compute and analyze different scenarios in all possible solutions in our cases of melanoma. In short, a reachability analysis in choroidal melanoma allows us to determine cancer cell proliferation and survival.

References

1. Calipel, A., Lefevre, G., Pouponnot, C., Mouriaux, F.: Mutation of B-Raf in human choroidal melanoma cells mediates cell proliferation and transformation through the MEK/ERK pathway. J. Biol. Chem. **278**(43), 42409–42418 (2003)
2. Clavel, M., Durán, F., Eker, S., Lincoln, P., Martí-Oliet, N., Meseguer, J., Talcott, C.L.: All About Maude - A High-Performance Logical Framework, How to Specify, Program and Verify Systems in Rewriting Logic, LNCS, vol. 4350. Springer (2007)
3. Delmas, C., Aragou, N., Poussard, S., Cottin, P., Darbon, J.M., Manenti, S.: MAP kinase-dependent degradation of p27Kip1 by calpains in choroidal melanoma cells requirement of p27Kip1 nuclear export. J. Biol. Chem. **278**(14), 12443–12451 (2003)
4. Hecquet, C., Lefevre, G., Valtink, M., Engelmann, K., Mascarelli, F.: cAMP inhibits the proliferation of retinal pigmented epithelial cells through the inhibition of ERK1/2 in a PKA-independent manner. Oncogene **21**(39), 6101 (2002)

5. Hwang, W., Hwang, Y., Lee, S., Lee, D.: Rule-based multi-scale simulation for drug effect pathway analysis. BMC Med. Inform. Decis. Mak. **13**(Suppl. 1), S4 (2013)
6. Krantz, B.A., Dave, N., Komatsubara, K.M., Marr, B.P., Carvajal, R.D.: Uveal melanoma: epidemiology, etiology, and treatment of primary disease. Clin. Ophthalmol. **11**, 279 (2017)
7. Lefevre, G., Calipel, A., Mouriaux, F., Hecquet, C.: Opposite long-term regulation of c-Myc and $p27^{Kip1}$ through overactivation of Raf-1 and the MEK/ERK module in proliferating human choroidal melanoma cells. Oncogene **22**(55), 8813 (2003)
8. Lo, J.A., Fisher, D.E.: The melanoma revolution: from UV carcinogenesis to a new era in therapeutics. Science **346**(6212), 945–949 (2014)
9. Manenti, S., Malecaze, F., Darbon, J.M.: The major myristoylated PKC substrate (MARCKS) is involved in cell spreading, tyrosine phosphorylation of paxillin, and focal contact formation. FEBS Lett. **419**(1), 95–98 (1997)
10. Martí-Oliet, N., Ölveczky, P.C., Talcott, C.L. (eds.): Logic, Rewriting, and Concurrency - Essays Dedicated to José Meseguer on the Occasion of his 65th Birthday. LNCS, vol. 9200. Springer, Heidelberg (2015)
11. Meseguer, J.: Conditional rewriting logic as a unified model of concurrency. Theor. Comput. Sci. **96**(1), 73–155 (1992)
12. Nieto, C., Escudero, F., Rivero, V., Hernández Galilea, E.: Choroidal metastasis from primary bone leiomyosarcoma. Int. Ophthalmol. **35**(5), 721–725 (2015)
13. Patel, M., Smyth, E., Chapman, P.B., Wolchok, J.D., Schwartz, G.K., Abramson, D.H., Carvajal, R.D.: Therapeutic implications of the emerging molecular biology of uveal melanoma. Clin. Cancer Res. **17**(8), 2087–2100 (2011)
14. Riesco, A., Santos-Buitrago, B., De Las Rivas, J., Knapp, M., Santos-García, G., Talcott, C.: Epidermal growth factor signaling towards proliferation: modeling and logic inference using forward and backward search. Biomed. Res. Int. **2017**, 11 (2017)
15. Santos-Buitrago, B., Riesco, A., Knapp, M., Alcantud, J.C.R., Santos-García, G., Talcott, C.: Soft set theory for decision making in computational biology under incomplete information. IEEE Access **7**, 18183–18193 (2019)
16. Santos-García, G., De Las Rivas, J., Talcott, C.L.: A logic computational framework to query dynamics on complex biological pathways. In: Saez-Rodriguez, J., et al. (eds.) 8th International Conference on Practical Applications of Computational Biology & Bioinformatics, PACBB 2014, 4–6 June 2014, Salamanca, Spain. AISC, vol. 294, pp. 207–214. Springer, Cham (2014)
17. Santos-García, G., Palomino, M., Verdejo, A.: Rewriting logic using strategies for neural networks: an implementation in Maude. In: Corchado, J.M., et al. (eds.) Proceedings of International Symposium on Distributed Computing and Artificial Intelligence, DCAI 2008, Salamanca, Spain, 22–24 October 2008. AISC, vol. 50, pp. 424–433. Springer, Berlin (2009)
18. Santos-García, G., Talcott, C.L., De Las Rivas, J.: Analysis of cellular proliferation and survival signaling by using two ligand/receptor systems modeled by Pathway Logic. In: Abate, A., Safránek, D., (eds.) Hybrid Systems Biology - Fourth International Workshop, HSB 2015, Madrid, Spain, 4–5 September 2015. Revised Selected Papers. LNCS, vol. 9271, pp. 226–245. Springer, Cham (2015)
19. Shtivelman, E., Davies, M.A., Hwu, P., Yang, J., Lotem, M., Oren, M.: Pathways and therapeutic targets in melanoma. Oncotarget **5**(7), 1701 (2014)
20. Talcott, C.L.: Pathway logic. In: Bernardo, M., Degano, P., Zavattaro, G., (eds.) SFM. LNCS, vol. 5016, pp. 21–53. Springer (2008)

21. Tenazinha, N., Vinga, S.: A survey on methods for modeling and analyzing integrated biological networks. IEEE/ACM Trans. Comput. Biol. Bioinform. **8**(4), 943–958 (2011)
22. Weng, G., Bhalla, U.S., Iyengar, R.: Complexity in biological signaling systems. Science **284**(5411), 92–96 (1999)

Predicting Promoters in Phage Genomes Using Machine Learning Models

Marta Sampaio, Miguel Rocha, Hugo Oliveira, and Oscar Dias(✉)

Centre of Biological Engineering, University of Minho, Braga, Portugal
{msampaio, odias}@ceb.uminho.pt, mrocha@di.uminho.pt,
hugooliveira@deb.uminho.pt

Abstract. The renewed interest in phages as antibacterial agents has led to the exponentially growing number of sequenced phage genomes. Therefore, the development of novel bioinformatics methods to automate and facilitate phage genome annotation is of utmost importance. The most difficult step of phage genome annotation is the identification of promoters. As the existing methods for predicting promoters are not well suited for phages, we used machine learning models for locating promoters in phage genomes. Several models were created, using different algorithms and datasets, which consisted of known phage promoter and non-promoter sequences. All models showed good performance, but the ANN model provided better results for the smaller dataset (92% of accuracy, 89% of precision and 87% of recall) and the SVM model returned better results for the larger dataset (93% of accuracy, 91% of precision and 80% of recall). Both models were applied to the genome of *Pseudomonas* phage phiPsa17 and were able to identify both types of promoters, host and phage, found in phage genomes.

Keywords: Machine learning · Genome analysis · Phages · Promoters

1 Introduction

Bacteriophages, or phages, are viruses that exclusively kill bacteria [1]. In the last decades, phages have been extensively studied and their genomic information has increased exponentially, mainly due to their therapeutic potential against bacterial infections, at a time when the rise of antibiotic resistance in pathogenic bacteria represents a serious health problem [2]. Thus, such abundance of data demands the development of bioinformatics methods to facilitate genome annotation. The main obstacle in genome annotation is the identification of promoters, which are specific DNA regions responsible for transcription initiation. Identification of promoters is difficult, because these are composed of short, non-conserved elements. However, it is crucial for understanding and characterising phage genetic regulatory networks, which may allow to design better phages with applications in biotechnology and medicine [3].

Promoters are poorly described for phage genomes. Indeed, only the phiSITE database provides a list of identified promoters for 29 phage genomes [4]. Some phages

F. Fdez-Riverola et al. (Eds.): PACBB 2019, AISC 1005, pp. 105–112, 2020.
https://doi.org/10.1007/978-3-030-23873-5_13

are able of encoding their own RNA polymerase (RNAP). Hence, besides host promoters, which are recognized by the host's RNAP, the genome of these phages contains promoters that are recognized by their intrinsic RNAP [5].

A few general-purpose promoter prediction tools for bacterial genomes have been developed, using diverse computational algorithms. The most recent tools are based on machine learning models, such as CNNpromoter_b, using deep learning networks [6], BPROM, using linear discriminant analysis (LDA) [7], and bTSS finder, using artificial neural networks (ANN) [8]. However, these tools still return numerous false positives [9]. Such tools were developed using bacterial promoters and only search for the typical bacterial motifs of the −35 and −10 elements (TTGACA and TATAAT, respectively), thus not being suitable for phages genomes. Therefore, these are not able to find promoters recognized by phage own RNAP nor host promoters with different motifs. Other tools, such as the widely-used PromoterHunter available on the phiSITE website, have additional disadvantages like requiring the weight matrices of the two promoter elements as input and limiting the size of the input genome sequence [4]. For phages, only PHIRE software was developed for predicting regulatory elements [10]. However, it only searches for conserved sequences with 20 base pairs or more and requires installation.

Therefore, in this work, machine learning models were trained using phage sequences for identifying both types of promoters found in phage genomes and different motifs of each promoter type.

2 Methods

2.1 Data

The positive data used to train the models was retrieved from the phiSITE database and available publications, consisting of 800 promoter sequences from 53 phage genomes. Since there are no sequences identified as non-promoters, sequence fragments of 65 base pairs were randomly selected from the 53 genomes to form the negative sets, provided that the selected fragment did not include a known promoter. There is no consensus length for promoters, so the fragment size was chosen according to the size of the biggest collected promoter.

As the number of promoters in the whole genome is several orders of magnitude lower than the non-promoters, having more negative than positive cases in the dataset should be more adequate for finding promoters in phage genomes. Therefore, two datasets were created, both comprising the 800 positive cases, though with different negative sets: Dataset1 including 1600 negative sequences and Dataset2 including 2400 negatives.

2.2 Features

Twelve different motifs were previously found in the collected phage promoters, using motif finder tools like MEME [11]. Eight of these motifs represent the elements of host promoters and four are recognized by the intrinsic RNAP (Table 1). Data features included the sizes and scores of these motifs, which were calculated using Position Specific Scoring Matrices (PSSM) with pseudocounts of 1, as well as information about phage lifecycle, family and host. The free energy value and the frequency of adenines and thymines were also calculated for each sequence, as these express the stability of the DNA molecule, which is expected to be lower in the promoter region. The free energy value was calculated by summing the unified nearest-neighbor (NN) energy values of each dinucleotide that were defined by SantaLucia et al. [12]. The datasets were standardized and the recursive feature elimination (RFE) method was used to select the most relevant features of the datasets. RFE was applied using Random Forests as estimator and removing one feature at each iteration. After applying this method, some features representing the host and motif sizes were eliminated from the datasets. The final datasets are available at: https://github.com/martaS95/PhagePromoter/Data.

Table 1. Description of the motifs identified in the collected data. In the consensus sequence, Y = C or T; M = A or C; R = A or G; W = A or T; N = A, C, G or T.

Type	Element	Size (bp)	Phages	Consensus sequence
Host	−10	6	Almost all (51)	TATAAT
Host	−35	6	Almost all (50)	TTGACA
Host	−10	8	T4 e CBB	TATAAATA
Host	−35	7	T4	GTTTACA
Host	−35	7	CBB	TGAAACG
Host	−35	9	T4	AWTGCTTTA
Host	−35	14	Lambda-like	$TTGCN_6TTGC$
Host	−35	14	Mu-like	CCATAACCCCGGAC
Phage	None	23	T7-like	TAATAAGACTCACTAAAGGGAGA
Phage	None	21	phi-C31	CCGGGTTGCCGACTCCCTTMC
Phage	None	27	phiKMV-like	CGACCCTGCCCTACTCCGGGCTYAAAT
Phage	None	32	KP34-like	AGCCTATAGCRTCCTACGGGGYGCTATGTGAA

2.3 Models

Machine learning models were built to classify sequence fragments as promoters or non-promoters, using four different models: artificial neural networks (ANN), support vector machines (SVM), random forests (RF) and k-nearest neighbors (KNN). For each algorithm, two models were trained with each dataset (Dataset1 or Dataset2), creating eight models. These models were optimized using the Grid Search method. Table 2 describes the tested values of model hyperparameters, which were selected empirically, and the best values obtained for the hyperparameters.

The models were further evaluated using cross-validation with 5-folds and the selected metrics were accuracy, precision and recall. Confusion matrices and Matthews correlation coefficients (MCCs) were also calculated for all models. These steps were performed using the Python library Scikit-learn [13].

Table 2. Hyperparameter values tested for each model, using Grid Search. The best values are highlighted by color: in red are the best values obtained using Dataset1 and in blue are the ones obtained using Dataset2. The values in green were the same for the two datasets.

Model	Parameter	Values tested for Grid Search
ANN	solver for weight optimization	lbfgs, sgd, **adam**
	activation function	identity, logistic, tanh, **relu**
	alpha	0.0001, **0.001**, **0.01**
	hidden layer size	**(15,)**,(25,),(50,),(75,),**(100,)**
SVM	C (regularization)	1,**2.26**,10,**15**,20
	gamma	**auto**,0.001,0.01, **0.05**, 0.1
	kernel	linear, **rbf**, poly, sigmoid
RF	number of trees in the forest	**20**, 40, 60, 80,**100**
	number of features	**auto**,2,3,6,10
	minimum number of samples to split an internal node	**2**, 3, **6**, 10
	minimum number of samples to be at a leaf node	**2**, 3, 6, 10
	maximum depth of the tree	2, 3, **None**
	bootstrap	**True**, **False**
	criterion	**gini**, **entropy**
KNN	Number of neighbors	3,**5**,**7**,9

3 Results and Discussion

The results of model evaluation are presented in Table 3. Globally, all models showed good results, as these presented high accuracy and precision and acceptable recall. The models trained with Dataset1 present higher recall while the models trained with Dataset2 have higher accuracy. The SVM and KNN models also show higher precision with Dataset2. For Dataset1, the model with best performance was the ANN model with 92% of accuracy, 89% of precision and 87% of recall. Although the RF model presented the highest precision (91%), it also had the lowest recall (83%). For Dataset2, both SVM and RF models presented the highest precision (91%) and the ANN model the highest recall (83%). The KNN model performed slightly worse for both datasets.

Since the differences between these metrics are not significant, confusion matrices were also generated to evaluate the performances of the models and MCCs were calculated from them. These results are represented in Tables 4, 5, 6 and 7.

As expected, the ANN model has the lowest number of false negatives (FN) and the highest number of false positives (FP) for both datasets. For Dataset1, the RF model has the lowest number of FP but the highest number of FN. For Dataset2, both RF and SVM models have the lowest number of FP, but the SVM has less FN than the RF model. Correlating these values using MCC, it is possible to see that for Dataset1, both SVM and ANN have the highest MCC value and SVM model has also the highest MCC value for Dataset2. Nevertheless, the small differences between the calculated MCCs indicate that all models present similar performances.

Table 3. Mean values of accuracy, precision and recall for each model after a 5-fold CV

	Dataset1			Dataset2		
Models	Accuracy	Precision	Recall	Accuracy	Precision	Recall
ANN	0.92	0.89	0.87	0.93	0.87	0.83
SVM	0.92	0.89	0.86	0.93	0.91	0.80
RF	0.92	0.91	0.83	0.93	0.91	0.79
KNN	0.91	0.89	0.84	0.92	0.90	0.78

Table 4. Confusion matrices of the ANN models for both datasets

Dataset1 (1600 negatives)

Real \ Predicted	**Positive**	**Negative**	**Total**
Positive	694	106	800
Negative	87	1513	1600
Total	781	1619	2400
MCC	0.82		

Dataset2 (2400 negatives)

Real \ Predicted	**Positive**	**Negative**	**Total**
Positive	661	139	800
Negative	94	2306	2400
Total	755	2445	3200
MCC	0.81		

Table 5. Confusion matrices of the SVM models for both datasets

Dataset1 (1600 negatives)

Real \ Predicted	**Positive**	**Negative**	**Total**
Positive	685	115	800
Negative	75	1525	1600
Total	760	1640	2400
MCC	0.82		

Dataset2 (2400 negatives)

Real \ Predicted	**Positive**	**Negative**	**Total**
Positive	641	159	800
Negative	68	2332	2400
Total	709	2491	3200
MCC	0.80		

Table 6. Confusion matrices of the RF models for both datasets

Dataset1 (1600 negatives)

Real \ Predicted	**Positive**	**Negative**	**Total**
Positive	666	134	800
Negative	69	1531	1600
Total	735	1665	2400
MCC	0.81		

Dataset2 (2400 negatives)

Real \ Predicted	**Positive**	**Negative**	**Total**
Positive	628	172	800
Negative	68	2332	2400
Total	696	2504	3200
MCC	0.79		

Table 7. Confusion matrices of the KNN models for both datasets

Dataset1 (1600 negatives)

Real \ Predicted	**Positive**	**Negative**	**Total**
Positive	672	128	800
Negative	81	1519	1600
Total	753	1647	2400
MCC	0.80		

Dataset2 (2400 negatives)

Real \ Predicted	**Positive**	**Negative**	**Total**
Positive	627	173	800
Negative	69	2331	2400
Total	696	2504	3200
MCC	0.79		

Data from confusion matrices confirms that the number of FN is higher than the number of FP in all models, which might be explained by the fact that the negative sequences selected to train the model were putative and not proven to be negative. Thus, the negative cases set may encompass promoter sequences that have not yet been identified which can prompt the models to predict a promoter as negative. Another possible explanation for these results is that some known promoters have motif sequences very distinct from the consensus, thus their scores regarding these features will be low, inducing the model to classify them as negatives.

Although all models presented similar performance, only two were selected to be applied to the case study: the ANN model trained with Dataset1 and the SVM model trained with Dataset2.

3.1 Case Study: *Pseudomonas* Phage PhiPsa17

The two models were applied to the genome of *Pseudomonas* phage phiPsa17, to test their predictive capacity. This lytic phage belongs to the *Podoviridae* family and was extracted from *Pseudomonas syringae*. It encodes its own RNAP which means its early genes are transcribed by the host RNAP whereas middle and late gene are transcribed by phage intrinsic RNAP [14]. Thus, two types of promoters are expected to be found in its genome: host promoters, with the −10 and/or −35 elements, and phage promoters with sequence similar to those of T7-like virus. There are no promoters of this phage in phiSITE. Analysing the study of Frampton et al. [14], 1 host promoter was identified in the early region of the genome using BPROM tool [7] and 11 phage promoters were identified using MEME [11], considering only the 100 base pairs upstream of the predicted genes. As all predicted genes are in the direct strand, the models were only applied to the direct strand of this genome and searched the whole genome sequence for promoters. The results predicted by both models are summarized in the Venn diagrams of Fig. 1.

The SVM model predicted 16 promoters, 3 host and 13 phage promoters, while the ANN predicted 25 promoters, 8 host and 17 phage promoters. 14 promoters were identified by both models (2 host and 12 phage promoters) and 12 correspond to the promoters previously predicted by Frampton et al. [14]. The other promoters predicted by the models have lower scores and less common motifs for the −35 and −10 elements. The promoters predicted by both models are close to the predicted genes, except for the 2 host promoters. As expected, the ANN model predicted more promoters than

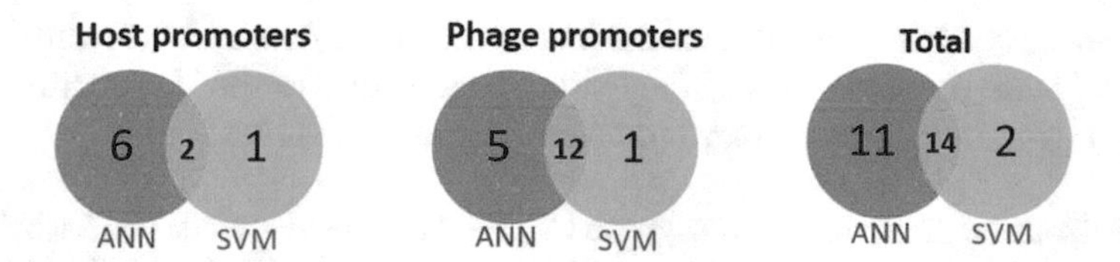

Fig. 1. Venn Diagrams representing the number of host and phage promoters predicted by SVM and ANN models for phiPsa17 phage genome

the SVM model. Thus, although a careful analysis of the results is required, these models presented good performance as both were able to identify all promoters predicted by Frampton et al. [14].

Comparing these results with the results of other tools, CNNpromoter_e predicted 31 host promoters for this genome, which is a higher number than expected for this genome since this phage only uses the host RNAP for transcribing early genes. None of these promoters were identified by the models. PromoterHunter tool only predicted three promoters which correspond to one host promoter predicted by both models and two predicted by the ANN model. Regarding phage promoters, PHIRE program identified 10 of the 11 phage promoters previously identified by MEME. Therefore, the developed models are better than these tools because they can recognize both promoter types, host and phage, and with different motifs of each, so there is no need to use different tools for identifying different promoters.

4 Conclusions

In this work, we propose methods to identify promoters in phage genomes, during genome annotation. Several machine learning models were trained with phage data, using two different datasets. All models showed good performance, but the ANN model provided better results for the smaller dataset whereas the SVM model returned better results for the larger dataset. The ANN model is expected to predict more promoters than the SVM model, so it can result in more false positives when applied to new data. On the other hand, the SVM predicts less promoters, so it may result in more false negatives.

The number of false negatives is higher than the number of false positives for all models, which might be explained by the high variety of phage promoter motifs and by the fact that the set of negative examples may encompass unidentified promoter sequences. In addition, the proportion between positive and negative cases in the datasets is much lower than the real proportion of promoters and non-promoters in a genome. The models identified phage promoters previously predicted by other tools and manually curated, but the number of phage genomes with identified promoters that were not used to train the models is very low. Therefore, having more phage promoter and non-promoter sequences experimentally identified is crucial to validate the models and improve promoter identification.

Nevertheless, as these models are the first to use phage data and to identify different motifs for both promoter types, they are undoubtedly useful for facilitating and speeding up the task of predicting promoters in phage genomes.

Acknowledgments. This study was supported by the Portuguese Foundation for Science and Technology (FCT) under the scope of the strategic funding of UID/BIO/04469/2019 unit and the Project POCI-01-0145-FEDER-029628. This work was also supported by BioTecNorte operation (NORTE-01-0145-FEDER-000004) funded by the European Regional Development Fund under the scope of Norte2020 - Programa Operacional Regional do Norte.

References

1. Salmond, G.P.C., Fineran, P.C.: A century of the phage: past, present and future. Nat. Rev. Microbiol. **13**(12), 777–786 (2015)
2. Haq, I.U., Chaudhry, W.N., Akhtar, M.N., Andleeb, S., Qadri, I.: Bacteriophages and their implications on future biotechnology: a review. Virol. J. **9**(1), 9 (2012)
3. Guzina, J., Djordjevic, M.: Bioinformatics as a first-line approach for understanding bacteriophage transcription. Bacteriophage **5**(3), e1062588 (2015)
4. Klucar, L., Stano, M., Hajduk, M.: phiSITE: database of gene regulation in bacteriophages. Nucleic Acids Res. **38**(Database issue), D366–D370 (2010)
5. Yang, H., Ma, Y., Wang, Y., Yang, H., Shen, W., Chen, X.: Transcription regulation mechanisms of bacteriophages: recent advances and future prospects. Bioengineered **5**(5), 300–304 (2014)
6. Umarov, R.K., Solovyev, V.V.: Recognition of prokaryotic and eukaryotic promoters using convolutional deep learning neural networks. PLoS ONE **12**(2), e0171410 (2017). https://doi.org/10.1371/journal.pone.0171410
7. Solovyev, V., Salamov, A.: Automatic annotation of microbial genomes and metagenomic sequences, January 2016
8. Shahmuradov, I.A., Mohamad Razali, R., Bougouffa, S., Radovanovic, A., Bajic, V.B.: bTSSfinder: a novel tool for the prediction of promoters in cyanobacteria and Escherichia coli. Bioinformatics **33**(3), 334–340 (2017)
9. Silva, S., Echeverrigaray, S.: Bacterial promoter features description and their application on e. coli in silico prediction and recognition approaches. In: Bioinformatics inTech, November 2012
10. Lavigne, R., Sun, W., Volckaert, G.: PHIRE, a deterministic approach to reveal regulatory elements in bacteriophage genomes. Bioinformatics **20**(5), 629–635 (2004)
11. Bailey, T.L., Williams, N., Misleh, C., Li, W.W.: MEME: discovering and analyzing DNA and protein sequence motifs. Nucleic Acids Res. **34**, W369–W373 (2006)
12. SantaLucia, J.: A unified view of polymer, dumbbell, and oligonucleotide DNA nearest-neighbor thermodynamics. Proc. Natl. Acad. Sci. **95**(4), 1460–1465 (1998)
13. scikit-learn: machine learning in Python — scikit-learn 0.21.2 documentation. https://scikit-learn.org/stable/index.html
14. Frampton, R.A., Acedo, E.L., Young, V.L., Chen, D., Tong, B., Taylor, C., Easingwood, R. A., Pitman, A.R., Kleffmann, T., Bostina, M., Fineran, P.C.: Genome, proteome and structure of a T7-Like bacteriophage of the kiwifruit canker phytopathogen pseudomonas syringae pv. actinidiae. Viruses **7**(7), 3361–3379 (2015)

Detection and Characterization of Local Inverted Repeats Regularities

Carlos A. C. Bastos[1,2(✉)], Vera Afreixo[1,3,4], João M. O. S. Rodrigues[1,2], and Armando J. Pinho[1,2]

[1] IEETA-Institute of Electronics and Informatics Engineering of Aveiro, Aveiro, Portugal
{cbastos,vera,jmr,ap}@ua.pt

[2] Department of Electronics, Telecommunications and Informatics, University of Aveiro, Aveiro, Portugal

[3] CIDMA-Center for Research and Development in Mathematics and Applications, Aveiro, Portugal

[4] Department of Mathematics, University of Aveiro, Aveiro, Portugal

Abstract. To explore the inverted repeats regularities along the genome sequences, we propose a sliding window method to extract the concentration scores of inverted repeats periodic regularities and the total mass of possible inverted repeats pairs. We apply the method to the human genome and locate the regions with the potential for the formation of large number of hairpin/cruciform structures. The number of found windows with periodic regularities is small and the patterns of occurrence are chromosome specific.

Keywords: Cruciform · Distance distribution · Inverted repeats · Periodic regularities

1 Introduction

Hairpin/cruciform structures are a type of non-B DNA structure with importance in biological processes and gene function [1]. DNA motifs that are known to potentially form non-B DNA structures are available at public databases [2,3]. Hairpins/cruciforms may form dynamically when certain conditions are met, such as the coiling state of DNA, but are less stable than the normal B-DNA conformation. Although their properties and relevance in several biological processes are acknowledged, evidence of their genomic location and mechanism of action are lacking *in vivo* [4,5].

The stem and loop lengths of hairpin/cruciforms structures seem to vary over a wide range. According to different authors, the stem lengths vary between 6 and 100 nucleotides, while loop lengths may range from 0 to 2000 nucleotides [2,6,7]. Shorter distances could favour the occurrence of these structures, but long distances have also been reported, such as the translocation breakpoints associated with human developmental diseases or infertility [4].

F. Fdez-Riverola et al. (Eds.): PACBB 2019, AISC 1005, pp. 113–120, 2020.
https://doi.org/10.1007/978-3-030-23873-5_14

The simultaneous occurrence of inverted repeats in a specific region are a required feature of local cruciform structures. However, some regions can greatly enhance the occurrence of hairpin/cruciforms conformations than others.

A DNA word analysis based on the distribution of the distances between adjacent symmetric words of length seven [8] showed a strong over-representation of distances up to 350, a feature that the authors considered might be associated with the potential for the occurrence of cruciform structures. Recently, the same research group extended their analysis to include distance distributions of non-adjacent inverted repeats, since adjacency is not a required condition for cruciform structures to form [9,10].

The present work focuses on identifying and characterising the local behaviour of inverted repeats. The occurrence of regular peaks and high mass in the cumulative distance distribution of symmetric word pairs will be explored.

2 Methods

This work aims to find, in the human genome, structures with regularity beyond the already well-known repetition structures published in the literature. Thus, we used pre-masked sequences available from the UCSC Genome Browser webpage [11]. These files contain the GRCh38 assembly sequences, with repeats reported by RepeatMasker [12] and Tandem Repeats Finder [13] masked with Ns.

Consider the alphabet $\mathcal{A} = \{A, C, G, T\}$ and let w be a symbolic sequence (word) defined in $\mathcal{A}^k$, where k is the length of w. The pair composed by one word, w, and the corresponding reversed complement word, w', is called an inverted repeats pair. For example, (ACT, AGT) is an inverted repeat pair.

In this work we analyse, along the human genome, the regularities in the distance distribution of inverted repeats by dividing the complete genome in successive windows containing 100 k nucleotides. Instead of separately analysing the distance distribution for each possible inverted repeat, as done in previous works [9,10], in the present work we analyse *cumulative* distance distributions of all possible inverted repeats. This keeps the data size manageable.

2.1 Distance Between Inverted Repeats

For all words of length k, we compute the frequency distributions of distances, f, between occurrences of each word and all succeeding reversed complements at distances between k and 4000.

For example, consider the sequence $ACTTTGTACTAAAGTTAAG$. Only four inverted repeats (w, w') of length $k = 3$ occur in this short sequence. The following lines show all occurrences of these inverted repeats, marked by underlines (w) and overlines (w'):

(ACT, AGT): $\underline{ACT}TTGT\underline{ACT}AA\overline{AGT}TAAG$,
(CTT, AAG): $A\underline{CTT}TGTACTA\overline{AAG}TT\overline{AAG}$,
(TTT, AAA): $AC\underline{TTT}GTACT\overline{AAA}GTTAAG$,

(TAA, TTA): $ACTTTGTAC\underline{TAA}AG\overline{TTA}AG$.

The previous sequence includes six distances to all the succeeding reversed complement words (distances: 12, 5, 10, 15, 8, and 5). Thus the cumulative distribution is $f(5) = 2, f(8) = f(10) = f(12) = f(15) = 1$ and $f(i) = 0$ for all other i values.

Motivated by previous work and considering the stem length of possible cruciform structures and considering computational limitations, we study words of length $k = 7$. For each word w we analyse distances up to 4000 nucleotides, but, if a N symbol is found, the search for w' is stopped, because the length of long stretches of Ns may be artificial.

2.2 Quantifying Periodic Regularities

In previous work, we detected words with strong periodic regularities in the complete human genome [10]. That work proposed a new measure for quantifying the periodic regularity of distributions, the *concentration score*. The proposed method is also able to find the fundamental period of the regularities.

The concentration score, s, for a given distribution $f(i)$, with $i = 1, 2, \ldots, N$ is computed in several steps [10]:

1. Obtain an auxiliary distribution g by sorting the frequencies in f in descending order.
2. Generate the family of *wrapped distributions* for f,
$$f_n(i) = \sum_{0 \leq j \leq \frac{N-i}{n}} f(i + jn), \text{ for } i = 1, 2, \ldots, n. \quad (1)$$
3. Generate the family of wrapped distributions, g_n, for g, using the previous procedure.
4. Compute the concentration score for each *wrapping period* n by the ratio
$$s(n) = \frac{\max f_n}{g_n(1)}. \quad (2)$$
5. The periodic regularity of distribution f is quantified by $S = \max s(n)$ and $P = \arg\max s(n)$ is its *fundamental period.*

Note that auxiliary distribution g eliminates any periodic regularity from f and that $\max(g_n) = g_n(1)$. Also, note that $s(n)$, S and P are not defined if $f(i) = 0$, for all i.

2.3 Number of Possible Pairs of Inverted Repeats

In a sequence, the total mass of the distribution f,

$$M = \sum_{i=k}^{4000} f(i), \quad (3)$$

corresponds to the number of possible pairs of inverted repeats occurring within a range of 4000 nt.

2.4 Windows Selection

In order to locate the sequence windows with the highest concentration scores or highest total mass we used quantile 0.999 as the discriminating threshold. This procedure resulted in the isolation of 30 windows with relevant concentration scores and 30 windows with relevant total mass.

3 Results

Table 1 shows the minimum, maximum and quartiles of concentration scores, S, and of total masses, M, over all the windows in each of the chromosomes and in the full human genome. The median of the set of S values over the complete genome is 1.50 and the inter-quartile range is 0.15 (Table 1), which shows that the majority of the windows have an S that shows low degree of periodic regularity. However, the maxima of S reveal that there are genomic regions that show high

Table 1. Order statistics of scores, S, and of total masses, M, over all the windows in each of the chromosomes and in the full human genome.

Concentration Score, S						Total mass, M					
chr	**min**	**Q1**	**med**	**Q3**	**max**	**chr**	**min**	**Q1**	**med**	**Q3**	**max**
1	1.00	1.43	1.50	1.58	4.16	**1**	1	2182	3053	4259	21361
2	1.16	1.42	1.48	1.55	3.56	**2**	15	2613	3517	4732	65495
3	1.21	1.42	1.48	1.56	10.88	**3**	137	2353	3273	4472	38696
4	1.20	1.40	1.47	1.55	6.06	**4**	119	2715	3723	4885	32671
5	1.19	1.42	1.48	1.56	5.69	**5**	42	2531	3586	4913	120365
6	1.21	1.42	1.48	1.55	2.85	**6**	250	2717	3602	4732	112782
7	1.00	1.42	1.48	1.55	4.47	**7**	1	2465	3573	4835	29132
8	1.19	1.42	1.49	1.56	3.00	**8**	3	2268	3259	4487	97380
9	1.19	1.43	1.50	1.57	7.83	**9**	135	2211	3101	4266	19694
10	1.00	1.43	1.48	1.56	3.35	**10**	1	2426	3234	4430	40031
11	1.00	1.43	1.50	1.57	3.79	**11**	1	2193	3030	4123	46349
12	1.00	1.43	1.50	1.56	5.68	**12**	1	2184	3107	4341	36793
13	1.17	1.40	1.46	1.53	5.53	**13**	539	3115	4183	5361	25881
14	1.16	1.43	1.48	1.56	9.43	**14**	135	2430	3374	4678	40828
15	1.00	1.43	1.50	1.56	2.39	**15**	1	2295	3128	4310	35519
16	1.25	1.44	1.50	1.58	4.00	**16**	26	1742	2661	3813	24775
17	1.00	1.43	1.50	1.57	5.92	**17**	1	2180	3114	4557	46507
18	1.00	1.42	1.47	1.54	4.03	**18**	1	2739	3683	4909	21117
19	1.00	1.46	1.55	1.67	7.51	**19**	1	1765	2655	3749	18806
20	1.28	1.44	1.50	1.58	4.79	**20**	2	1801	2507	3729	26786
21	1.00	1.42	1.48	1.55	3.00	**21**	1	2840	3942	5171	18180
22	1.28	1.44	1.50	1.59	4.00	**22**	11	1781	2856	4810	21097
X	1.00	1.47	1.56	1.67	11.70	**X**	1	1595	2412	3513	72594
Y	1.00	1.46	1.55	1.67	11.70	**Y**	1	1604	2727	4229	41404
all	1.00	**1.42**	**1.50**	**1.57**	11.70	**all**	1	2312	3278	4551	120365

degree of periodic regularity. The range of fundamental periods found is 30–102 showing diversity on the period regularities (see Table 2).

The selection criteria defined in Sect. 2.4 results in a threshold of 4.09 for the concentration score and a threshold of 27519 for the total mass. Using the criteria we observed that the occurrence of windows with high periodic regularity is heterogeneous among the chromosomes. Table 2 shows the windows with the highest concentration scores. Only 14 chromosomes contain regions with high periodic regularity. Figure 1 shows, as an example, the behaviour of the concentration score and of the total mass along the X chromosome. There are 5 windows with

Table 2. Windows with the 0.1% highest concentration scores and the windows with the 0.1% highest total mass.

Highest S

chr	win #	S	M	P
X	2	11.70	41404	61
Y	2	11.70	41404	61
3	1958	10.88	14845	48
X	3	10.02	18395	61
Y	3	10.02	18395	61
X	1	9.46	27519	61
Y	1	9.46	27519	61
14	1053	9.43	24554	**102**
9	4	7.83	4594	61
X	6	7.70	8879	44
Y	6	7.70	8879	44
19	2Y	7.51	3481	84
X	X	6.21	3786	40
Y	X	6.21	3786	40
4	1865	6.06	5547	28
17	10	5.92	2886	57
17	3	5.69	5182	45
5	4	5.69	14260	62
12	504	5.68	2744	36
5	1781	5.65	6145	43
13	11X	5.53	17242	34
4	1813	5.18	7638	51
13	11Y	4.84	10887	34
20	605	4.79	4145	46
5	1811	4.65	180X	33
12	405	4.48	16585	**30**
7	1547	4.47	18310	44
19	228	4.33	3348	84
19	234	4.32	2670	84
1	2368	4.16	5844	49

Highest M

chr	win #	S	M
5	674	1.39	120365
6	315	1.23	112782
8	1295	1.79	97380
6	1609	1.26	73271
X	531	1.76	72594
8	576	1.57	72053
2	685	1.16	65495
X	91	1.27	60390
X	1157	1.62	50189
17	135	1.48	46507
11	1147	1.88	46349
2	1948	1.97	44469
X	2	11.70	41404
Y	2	11.70	41404
5	1262	2.13	41007
14	859	2.27	40828
10	575	2.80	40031
3	1888	1.81	38696
12	866	2.14	36793
Y	2X	1.78	36129
2	439	1.47	35951
15	231	1.52	35519
2	2265	2.56	34891
6	1703	2.44	32718
4	1763	1.72	32671
3	1119	2.19	31790
8	353	1.24	31471
8	38	1.33	30288
7	581	1.31	29132
15	253	1.58	27826

win #- window number in chromosome

S- concentration score; M- total mass; P- fundamental period;

S above the threshold and 4 windows with M above the threshold, only one of the identified windows is common to both criteria. The scores S and M for the set of all the windows in the human genome are weakly correlated, Pearson's correlation coefficient is 0.29.

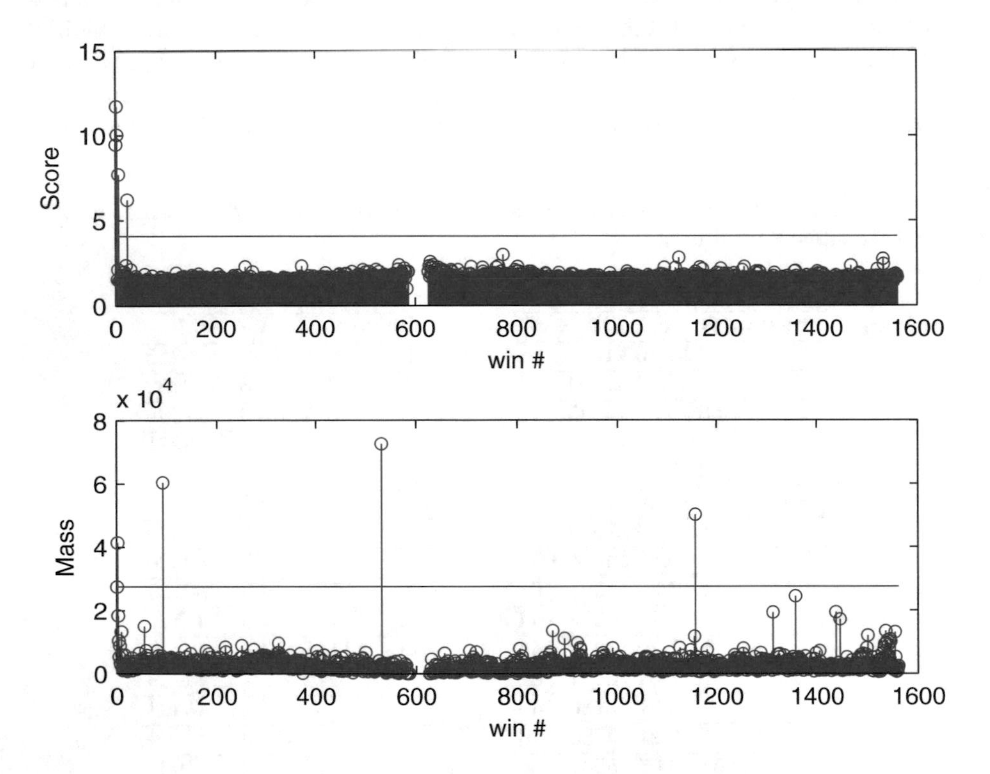

Fig. 1. The concentration score and the total mass of possible inverted repeats pairs, measured in 100 knt windows along chrX. The horizontal red lines correspond to the thresholds obtained with the windows selection criteria. Gaps around window 600 are a consequence of long stretches of the symbol N in the sequence.

Figures 2 and 3 show the cumulative distance distributions of, respectively, the window with the highest concentration score and the window with the highest total mass of possible inverted repeats pairs.

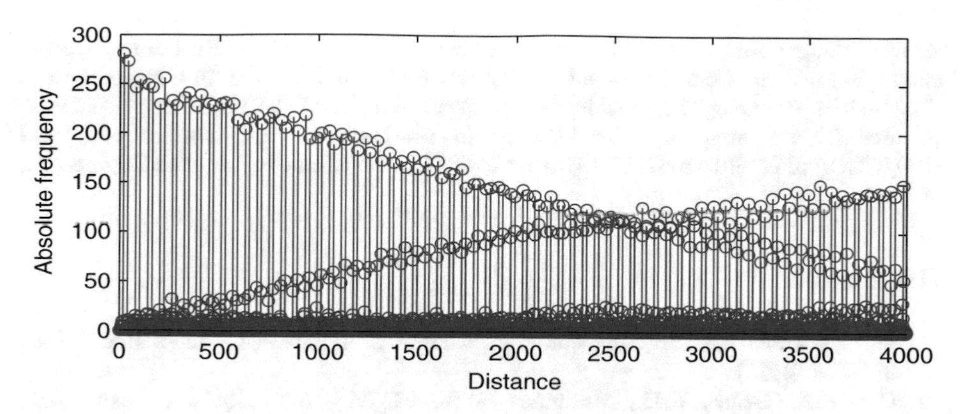

Fig. 2. Cumulative distance distribution of the window with the highest concentration score ($S = 11.70$) in chromosome X (chrX:100001–200000).

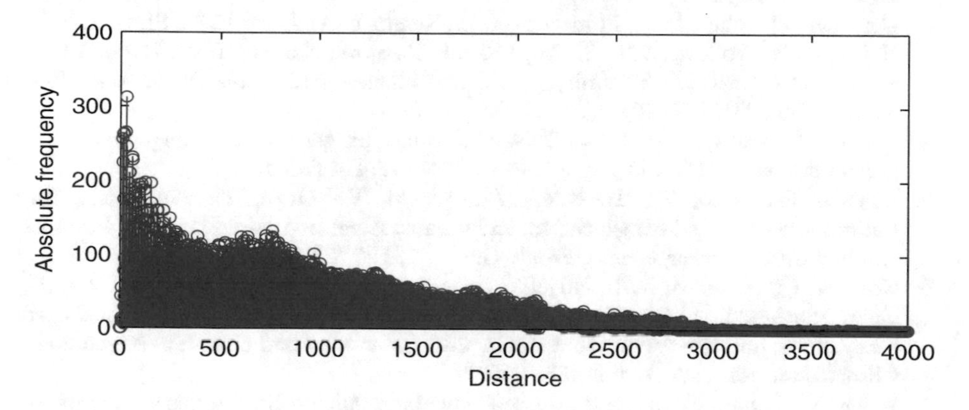

Fig. 3. Cumulative distance distribution of the window with the highest number of possible inverted repeat pairs ($M = 120365$) in chromosome 5 (chr5:67300001–67400000).

4 Discussion and Conclusion

Motivated by the potential connection between the occurrence of inverted repeats pairs with the possible formation of hairpin/cruciform structures, we explored the behaviour of the inverted repeats pairs in terms of periodic regularity of its occurrence and the total mass.

We identified genomic regions with atypically high score values indicative of the frequent occurrence of inverted repeats pairs at regular intervals. We also identified regions with a large number of possible inverted repeats pairs.

The patterns of periodic regularities and of total mass seem to be specific for each chromosome. The only exceptions are the patterns for the X and Y chromosomes, which are partially similar, as expected since they share parts of their genomic sequences.

Acknowledgment. This work was supported by FEDER ("Programa Operacional Fatores de Competitividade" - COMPETE) and FCT ("Fundação para a Ciência e a Tecnologia"), within the projects UID/MAT/04106/2019 to CIDMA (Center for Research and Development in Mathematics and Applications) and UID/CEC/00127/2019 to IEETA (Institute of Electronics and Informatics Engineering of Aveiro).

References

1. Du, Y., Zhou, X.: Targeting non-B-form DNA in living cells. Chem. Rec. **13**(4), 371–384 (2013)
2. Cer, R.Z., Bruce, K.H., Mudunuri, U.S., Yi, M., Volfovsky, N., Luke, B.T., Bacolla, A., Collins, J.R., Stephens, R.M.: Non-B DB: a database of predicted non-B DNA-forming motifs in mammalian genomes. Nucl. Acids Res. **39**(suppl–1), D383–D391 (2010)
3. Cer, R.Z., Donohue, D.E., Mudunuri, U.S., Temiz, N.A., Loss, M.A., Starner, N.J., Halusa, G.N., Volfovsky, N., Yi, M., Luke, B.T., et al.: Non-B DB v2. 0: a database of predicted non-B DNA-forming motifs and its associated tools. Nucl. Acids Res. **41**(D1), D94–D100 (2012)
4. Bacolla, A., Wells, R.D.: Non-B DNA conformations, genomic rearrangements, and human disease. J. Biol. Chem. **279**(46), 47411–47414 (2004)
5. Inagaki, H., Kato, T., Tsutsumi, M., Ouchi, Y., Ohye, T., Kurahashi, H.: Palindrome-mediated translocations in humans: A new mechanistic model for gross chromosomal rearrangements. Front. Genet. **7**, 125 (2016)
6. Kolb, J., Chuzhanova, N.A., Högel, J., Vasquez, K.M., Cooper, D.N., Bacolla, A., Kehrer-Sawatzki, H.: Cruciform-forming inverted repeats appear to have mediated many of the microinversions that distinguish the human and chimpanzee genomes. Chromosom. Res. **17**(4), 469–483 (2009)
7. Wang, Y., Leung, F.C.: Long inverted repeats in eukaryotic genomes: recombinogenic motifs determine genomic plasticity. FEBS Lett. **580**(5), 1277–1284 (2006)
8. Tavares, A.H., Pinho, A.J., Silva, R.M., Rodrigues, J.M., Bastos, C.A., Ferreira, P.J., Afreixo, V.: DNA word analysis based on the distribution of the distances between symmetric words. Sci. Rep. **7**(1), 728 (2017)
9. Bastos, C.A.C., Afreixo, V., Rodrigues, J.M.O.S., Pinho, A.J.: An analysis of symmetric words in human DNA: adjacent vs non-adjacent word distances. In: PACBB 2018 - 12th International Conference on Practical Applications of Computational Biology & Bioinformatics, Toledo, Spain, June 2018
10. Bastos, C.A.C., Afreixo, V., Rodrigues, J.M.O.S., Pinho, A.J., Silva, R.: Distribution of distances between symmetric words in the human genome: Analysis of regular peaks. Computational Life Sciences, Interdisciplinary Sciences (2019)
11. Kent, W., Sugnet, C., Furey, T., Roskin, K., Pringle, T., Zahler, A., Haussler, D.: The human genome browser at UCSC. Genome Res. **12**(6), 996–1006 (2002)
12. Smit, A.F.A., Hubley, R., Green, P.: RepeatMasker Open- 4.0 (2013–2015)
13. Benson, G.: Tandem repeats finder: a program to analyze dna sequences. Nucl. Acids Res. **27**(2), 573 (1999)

An Efficient and User-Friendly Implementation of the Founder Analysis Methodology

Daniel Vieira[1(✉)], Mafalda Almeida[1], Martin B. Richards[2], and Pedro Soares[1,3]

[1] CBMA (Centre of Molecular and Environmental Biology), Department of Biology, University of Minho, Campus de Gualtar, 4710-057 Braga, Portugal
jdanav@gmail.com, sfsmafalmeida@gmail.com
[2] Department of Biological Sciences, School of Applied Sciences, University of Huddersfield, Queensgate, Huddersfield HD1 3DH, UK
m.b.richards@hud.ac.uk
[3] Institute of Science and Innovation for Bio-Sustainability (IB-S), University of Minho, Campus de Gualtar, 4710-057 Braga, Portugal
pedrosoares@bio.uminho.pt

Abstract. Founder analysis is a sophisticated application of phylogeographic analysis. It comprises the estimation of timing and impact of migrations in current populations by taking advantage of the non-recombining property of certain marker systems (in the first instance, mitochondrial DNA) and therefore the possibility of building realistic phylogenetic trees. Given two populations, a source and a sink, and an assumed direction of migration between them, we can identify founder haplotypes, date the founder clusters deriving from them, and also estimate the proportions of lineages in each migration event. Despite being a methodology dating back nearly two decades and having been featured in numerous research articles, its use has mostly been restricted to a handful of research groups due to the cumbersome and time-consuming calculations it entails, a hindrance which stems not only from the often prohibitively large volume of data being dealt with but also the intricacies involved in the detection of founders. We have developed a Python-based tool with a user-friendly interface in response to these issues, providing a fast, automatized approach to the founder analysis pipeline and additional useful features within this context, expediting this step efficiently and allowing more hypotheses to be tested in a reliable and easily reproducible manner.

Keywords: Phylogenetics · Haplotype · Mitochondrial DNA · Population genetics

1 Introduction

Founder analysis is a methodology established in archaeogenetics nearly two decades ago that aims to establish the most probable times of migration between two populations or geographic areas and the proportion of lineages in the present-day sink population that traces back to each migratory event [1]. At its core, founder analysis is a

F. Fdez-Riverola et al. (Eds.): PACBB 2019, AISC 1005, pp. 121–128, 2020.
https://doi.org/10.1007/978-3-030-23873-5_15

quantitative approach to phylogeographic analysis [2, 3]; it requires a phylogenetic reconstruction of at least two groups (source and sink), a molecular clock in order to estimate a time frame for the events and an established hypothesis regarding the hypothetical demographic relationships between these two groups; the hypothesis is commonly obtained or supported from other scientific areas such as archaeology, paleoclimatology, paleontology, anthropology, geology and linguistics (Fig. 1).

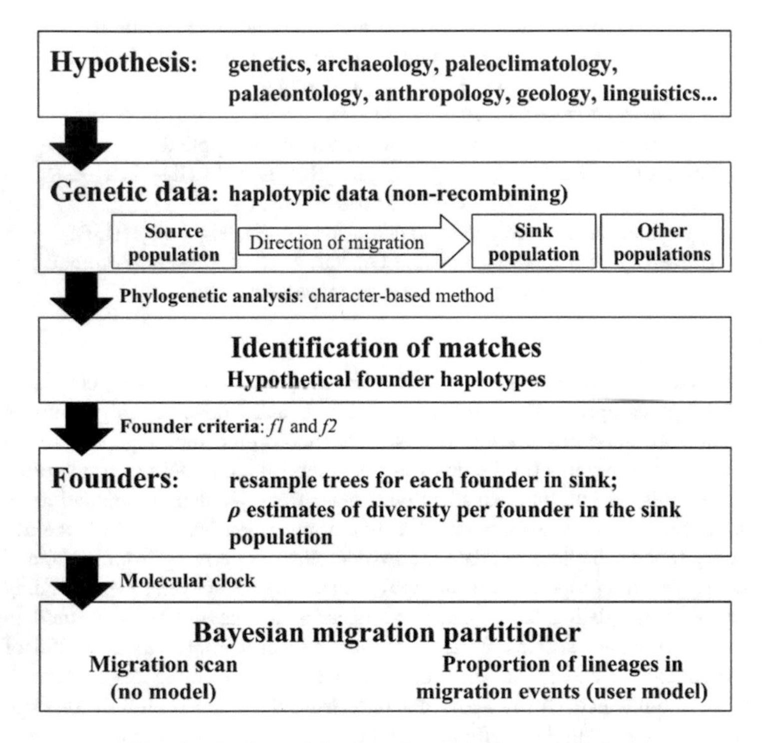

Fig. 1. Outline of the founder analysis methodology

A phylogenetic reconstruction for a founder analysis requires a character-based methodology in the tree building, since each node needs to be exactly assigned to a specific haplotype. Shared haplotypes between populations in the phylogeny (direct matches or inferred matches) are potential founders as they are evidence of gene flow between the two studied populations. The founder analysis will be performed considering a model that establishes a directionality of the migration, from a source population to a recipient population, commonly labelled source and sink respectively. The diversity accumulated within a founder clade in the sink population can be converted using a molecular clock for an estimate of the arrival of that clade in that region or population using the ρ estimate [4, 5].

Nevertheless, not all matches necessarily represent founders. In many genetic systems, for example in mtDNA, there is a high chance of homoplasy (the same mutation occurring twice independently) in the source and sink populations. Also, very recent migrations, which are irrelevant for the prehistoric migrations normally under investigation, will generate direct matches between populations. In order to decrease the skewing effect of both situations, two levels of stringency of the analysis have been introduced, called the *f1* and *f2* criteria [1]. The stringency criterion imposes the rule that a matching haplotype needs to display some level of deeper ancestry in the source population so that it can be labelled a founder, requiring that at least one derived branch of the founder haplotype (*f1* criterion) or two derived branches (*f2* criterion) are detected in the source population [1] (Fig. 2).

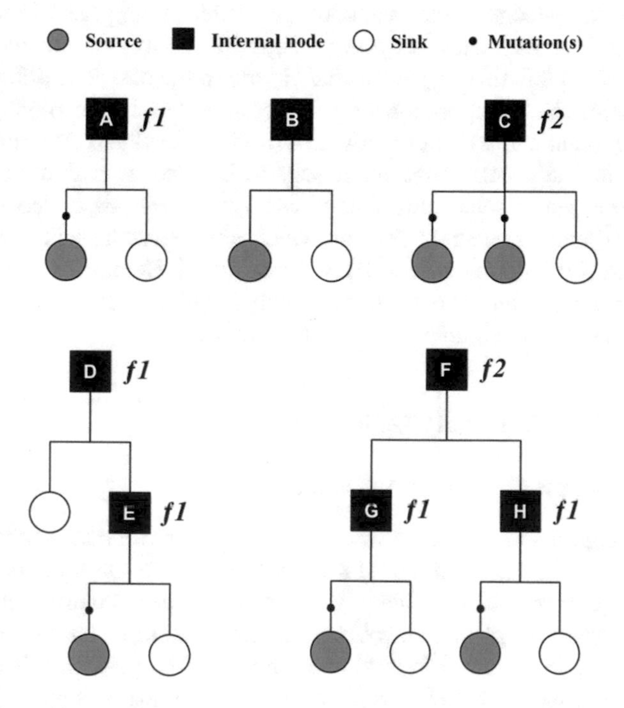

Fig. 2. Examples of internal node classifications upon applying the *f1* and *f2* criteria on their respective subtrees. Note that the absence of mutations in the source population under internal node B preclude it from being considered a founder.

Following the identification of all the founders in the phylogeny under a given criterion, we need to resample the subtrees of the founder clades in the sink population only, to assess the diversity and obtain individual age estimates from each subtree. Each founder will be statistically allocated to the different migration events stipulated in the model based on its age estimate and corresponding standard error of the ρ estimate [6, 7] in a Bayesian framework [1]. One later modification of the methodology was the possibility to perform a migratory scan without any prior model [7]. This does not represent any change in the overall procedure, but simply the stipulation of equally-spaced migration events (of 100 or 200 years).

Although results from founder analyses have been cited hundreds of times in the literature, the application of the method to phylogenetic data has been quite scarce. In fact, all articles to date employing a formal application, using the defined criteria described above, are at least co-authored by one of two present authors [1, 7–15]. A likely reason for this scenario is the fact that the methodology is very labour intensive and, most importantly, requires an appreciation of the principles of founder analysis, mainly at the level of selection of founders according to the defined criteria.

The best solution for this issue is the development of user-friendly software that not only accelerates processes that would traditionally take weeks but also reduces the required expert knowledge in the methodology. While some parts of the analyses are trivial to code (age estimates, Bayesian migration analysis), the identification of founders under both criteria has proven challenging in the past. The difficulty increases substantially with the incorporation of other populations in the phylogeny (beyond source and sink populations). These provide a more detailed and accurate phylogenetic reconstruction and allow the same phylogeny to be used in multiple models (where these populations can be source and sink populations), while simultaneously providing a much more complex tree structure from which any algorithm will need to resample only source and sink information for the different nodes. Herein we therefore present a newly developed algorithm[1] capable of accurately identifying founder nodes (under the *f1* and *f2* criteria) within complex phylogenetic trees.

2 Founder Analysis Software

2.1 Development and Main Functionalities

We initially programmed the latest working build of the founder analysis (shortened FA) software in Python 2.7 and ported it to Python 3 to ensure long-term support; we limited the usage of modules outside of the Python standard library to **Matplotlib**, for the output of interactive plots, and **py2exe** for the creation of binary executable files, tested successfully in current Windows and Unix operating systems. We designed the graphical user interface using **Tkinter**; it aims to display the hierarchical structure of a mtDNA tree alongside statistical data of interest.

As initial input, FA accepts an XML file containing a single mtDNA tree of any size, provided its format matches that of the complete tree available for download on the PhyloTree platform [16]. Each node element must include the attributes Node ID and HG. Although the existence of additional attributes is acceptable, they will be discarded. The value of the Node ID attribute must be unique for each element; detection of duplicates in the XML file produces a warning indicating the line in which the anomaly was found, and the parsing is aborted. Conversely, the HG attribute may hold zero or more values, repeatable across different nodes, separated by a comma, space or semi-colon (or any combination of the three).

[1] http://github.com/jdanav/FA.

```
<Node ID="Internal_node" HG="mutation_1, mutation_2">
  <Node ID="Leaf_node" HG="" />
</Node>
```

2.2 Data Input and Basic Analysis

Upon successful parsing of an input XML file, two types of objects are created:

- a Node object for each element of the tree, containing information such as the IDs of its parent and offspring nodes, the mutations read from the HG attribute, and a four-character code denoting whether it is an internal (haplogroup) or terminal (sample) node;
- a Tree object, comprising a dictionary which preserves the hierarchy of the input file and a collection of every subtree within it, one for each internal node.

The ρ estimate is then computed for each subtree, as per Eq. 1 [4]:

$$\rho = (m_1 + m_2 + \ldots + m_n)/n \tag{1}$$

where m_i corresponds to the number of mutations found along the path connecting the ith terminal node to the root of the subtree, and n the total number of terminal nodes. The age estimate (AE) is calculated according to Eq. 2 [17]:

$$AE = exp(-exp(-0.0263 \times (\rho + 40.2789))) \times \rho \times 3624 \tag{2}$$

Standard errors and confidence intervals are subsequently calculated based on the values of ρ and AE, respectively, concluding the first set of estimations. At this point, the user may then provide additional files assigning the status of source or sink to the terminal nodes of the original tree. These files consist of two tab-separated columns, the first of which corresponding to the node ID and the second to the labels "source" or "sink" (any capitalization for these tags is accepted); nodes which are not included in this file, or are otherwise followed by a different label, will be marked as "undefined" and disregarded for the purpose of founder and migration analyses.

2.3 Detection of Founders and Estimation of Migration Contributions

The first step of the founder analysis starts in the level furthest from the root of the tree, with the assessment of the terminal nodes' user-provided status; the parent node's suitability as a founder is then evaluated using both the *f1* and *f2* criteria. The process is repeated for any terminal nodes at its level, if applicable, and continues upwards towards the root until all nodes have been updated. The value of ρ is re-estimated for each criterion taking only sink nodes into account. Similarly, new values of AE, standard errors and confidence intervals are computed. Note that at this point *f1* and *f2* are being analyzed concurrently, such that there are three possible values for any given node. Within the FA software these are displayed in separate tabs, designated "Main statistics," "*f1* statistics" and "*f2* statistics." Providing a new file with different labels for the terminal nodes will reset the latter two tabs.

Migration clusters can be computed at this point. The user is prompted to provide a mutation rate and either a range of dates and an interval or a custom set of dates; options to exclude singletons (i.e. founders with only one terminal node) and to use an adjusted value for the number of leaves (effective number, used for a more correct estimation of uncertainty [7]) are also given. The individual proportions of the founders for each of the given dates are calculated; mean contributions of each migration date and deviations from the mean are also estimated. As with the previous analysis, these results are shown in separate tabs, one for each criterion, displaying only the relevant founders. Interactive stacked bar plots for the individual proportions, or plots of the mean contribution and deviations of the migrations, can be generated.

2.4 Additional Functionalities

Results shown in any given tab may be exported as tab-separated text files for processing in external software, if desired. Plots are customizable upon creation and may also be exported to PNG format.

The list of samples (terminal nodes) in a tree can easily be extracted, and a function to produce a Newick string representing the original tree structure is also provided.

We added a random XML tree generator so that an example template with a user-specified number of nodes may be created. This feature was useful during early stages of development as to assess how the size of a tree would affect the software performance.

2.5 Case Study: The Expansion of the Polynesian Motif into Remote Oceania

To test the FA tool, a simple case was investigated. The mitochondrial haplogroup B4a1a1, often called the Polynesian motif due to its high frequency in the Polynesian Islands, was the major haplogroup carried by the first settlers of the islands of Remote Oceania [18], an expansion that started as early as 3300–3100 years with the settlement of Vanuatu [19]. As source, we considered the sampled populations in the region called Near Oceania (including New Guinea) and, as sink the sampled individuals from Vanuatu to Hawaii (Remote Oceania). The B4a1a XML tree contained 605 nodes and 2015 leaves, from which 754 were from the stipulated source and 332 were from the sink. We detected 31 founders with the f1 criterion and 21 with the f2 criterion. A mutation rate average of 2590 years for a mutation to happen was considered for this time range. We used the software tool to perform two analyses: in one a migration scan from 0 to 10,000 years spaced by intervals of 100 years was considered, and on a second one performed considering two migration points, one for the first settlement at 3100 years and one for more recent gene flow at 900 years (obtained from the previous scan).

Figure 3 displays the results obtained for both the f1 and f2 criteria, with graphics provided by the tool. Figure 3A shows two peak of probable migration times, one as

3400 or 3500 years (in criteria *f1* and *f2*, respectively), not far from the expected initial time of settlement of Remote Oceania [19], and a second peak appears at 900 years. As a second analysis we considered two migration times, one at 3100 years (more adjusted to the archaeological data) and one at 900 years (as obtained from the scan) and we estimated the proportion of lineages involved in each migration (Fig. 3B), corresponding to about 80% of lineages in the first migrations and 20% in the second using both criteria. The tool also generates a graphic on the statistical allocation of each founder to a given migration (not shown). Although the analysis is crude and would require further refinement, we consider that it displays the potential of the tool under development.

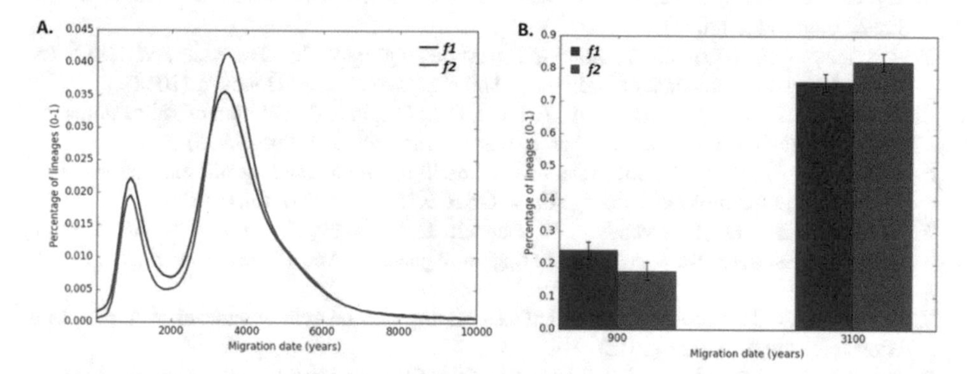

Fig. 3. Founders analysis results for a migration model from Near to Remote Oceania as generated by the FA tool under development; **A.** a migration scan of intervals of 100 years; **B.** a two-migration model at 3100 and 900 years.

3 Concluding Remarks

Although there is room for improvement, in terms of memory usage and execution times for the founder detection step, the software at this stage is capable of correctly performing the estimations described herein in a fraction of the time it would take had they been done manually. User feedback has been invaluable to provide insight into issues in need of fine-tuning or implementing desirable features. In addition, the complexity of the phylogenetic trees, comprising not only lineages from a source and a sink population but also several other leaves intermixed with those from the populations of interest, has led to a constant process of adaptation and improvement in the code following inspection by specialist users.

We hope that the future implementation of the founder analysis algorithm into an accessible and intuitive tool can lead to a more widespread use of this useful exploratory methodology, ultimately enriching archaeogenetics and studies of migration and dispersal more widely.

Acknowledgements. D.V., P.S. and M.B.R. acknowledge FCT support through project PTDC/EPH-ARQ/4164/2014 partially funded by FEDER funds (COMPETE 2020 project 016899). PS is supported by FCT, European Social Fund, Programa Operacional Potencial Humano and the FCT Investigator Programme and acknowledges CBMA strategic programme UID/BIA/04050/2019 funded by national funds through the FCT I.P. M.B.R. received support from a Leverhulme Trust Doctoral Scholarship programme.

References

1. Richards, M., et al.: Tracing European founder lineages in the Near Eastern mtDNA pool. Am. J. Hum. Genet. **67**, 1251–1276 (2000)
2. Avise, J.C.: Phylogeography: The History and Formation of Species. Harvard University Press, Cambridge (2001)
3. Richards, M.B., Torroni, A., Soares, P.: Phylogeography. In: Trevathan, W. (ed.) The International Encyclopedia of Biological Anthropology. Wiley, Hoboken (2018)
4. Forster, P., Harding, R., Torroni, A., Bandelt, H.J.: Origin and evolution of native American mtDNA variation: a reappraisal. Am. J. Hum. Genet. **59**, 935–945 (1996)
5. Macaulay, V., Soares, P., Richards, M.B.: Rectifying long-standing misconceptions about the ρ statistic for molecular dating. PLoS ONE **14**(2), e0212311 (2019)
6. Saillard, J., Forster, P., Lynnerup, N., Bandelt, H.J., Nørby, S.: mtDNA variation among Greenland Eskimos: the edge of the Beringian expansion. Am. J. Hum. Genet. **67**, 718–726 (2000)
7. Soares, P., et al.: The expansion of mtDNA haplogroup L3 within and out of Africa. Mol. Biol. Evol. **29**, 915–927 (2012)
8. Brandão, A., et al.: Quantifying the legacy of the Chinese Neolithic on the maternal genetic heritage of Taiwan and Island Southeast Asia. Hum. Genet. **135**, 363–376 (2016)
9. Gandini, F., et al.: Mapping human dispersals into the Horn of Africa from Arabian Ice Age refugia using mitogenomes. Sci. Rep. **6**, 25472 (2016)
10. Hernández, C.L., et al.: Early Holocenic and Historic mtDNA African signatures in the Iberian Peninsula: the Andalusian region as a paradigm. PLoS ONE **10**, e0139784 (2015)
11. Pereira, J.B., et al.: Reconciling evidence from ancient and contemporary genomes: a major source for the European Neolithic within Mediterranean Europe. Proc. Biol. Sci. **284**, 20161976 (2017)
12. Rito, T., et al.: The first modern human dispersals across Africa. PLoS ONE **8**, e80031 (2013)
13. Soares, P., Rito, T., Pereira, L., Richards, M.B.: A genetic perspective on African prehistory. In: Jones, C.S., Stewart, A.B. (eds.) Africa from MIS 6-2: Population Dynamics and Paleoenvironments. Springer, Dordrecht (2016)
14. Soares, P., et al.: Resolving the ancestry of austronesian-speaking populations. Hum. Genet. **135**, 309–326 (2016)
15. Al-Abri, A., et al.: Pleistocene-Holocene boundary in Southern Arabia from the perspective of human mtDNA variation. Am. J. Phys. Anthropol. **149**, 291–298 (2012)
16. van Oven, M., Kayser, M.: Updated comprehensive phylogenetic tree of global human mitochondrial DNA variation. Hum. Mutat. **30**(2), E386–E394 (2009)
17. Soares, P., et al.: Correcting for purifying selection: an improved human mitochondrial molecular clock. Am. J. Hum. Genet. **84**, 740–759 (2009)
18. Soares, P., et al.: Ancient voyaging and Polynesian origins. Am. J. Hum. Genet. **88**(2), 239–247 (2011)
19. Bedford, S., Spriggs, M., Regenvanu, R.: The Teouma Lapita site and the early human settlement of the Pacific Islands. Antiquity **80**(310), 812–828 (2016)

Visualization of Similar Primer and Adapter Sequences in Assembled Archaeal Genomes

Diogo Pratas[1(✉)], Morteza Hosseini[1], and Armando J. Pinho[1,2]

[1] IEETA, University of Aveiro, Aveiro, Portugal
{pratas,seyedmorteza,ap}@ua.pt
[2] DETI, University of Aveiro, Aveiro, Portugal

Abstract. Primer and adapter sequences are *synthetic* DNA or RNA oligonucleotides used in the process of amplification and sequencing. In theory, while similar primer sequences can be present on assembled genomes, adapter sequences should be trimmed (filtered) and, hence, absent from assembled genomes. However, given ambiguity problems, inefficient parameterization of trimming tools, and others, uncommonly they can be found in assembled genomes, on an exact or approximate state. In this paper, we investigate the occurrence of exact and approximate primer-adapter subsequences in assembled and, specifically, in the whole archaeal genomes of the NCBI database. We present a new method that combines data compression with custom signal processing operations, namely filtering and segmentation, to localize and visualize these regions given a defined similarity threshold. The program is freely available, under GPLv3 license, at https://github.com/pratas/maple.

Keywords: Genomics · Primer-adapter sequences · Approximate search · Data compression · Visualization · Signal processing · Archaeal genomes

1 Introduction

Primer and adapter sequences are *synthetic* DNA/RNA oligonucleotides used in the process of amplification and sequencing. Primer sequences, usually with lengths between 12 and 30 nucleotides, are used to prime DNA replication reactions, namely to amplify specific DNA sequences and obtain an amplicon [1]. Primers are usually designed as a pair, the forward and reverse primer, while for sequencing a single primer can be used. Adapters, also with short lengths, are used for *fishing* a (generally unknown) DNA sequence of interest. Adapters are a key component of the next-generation sequencing (NGS) workflow, namely for improving sample multiplexing.

Depending on experimental protocols and bioinformatics tools, the occurrence of primer-adapter sequences in raw sequence reads, along with endogenous and exogenous sources, is expected. While primer sequences may have similarity

F. Fdez-Riverola et al. (Eds.): PACBB 2019, AISC 1005, pp. 129–136, 2020.
https://doi.org/10.1007/978-3-030-23873-5_16

according to specific DNA sequences, adapter sequences should not contain high similarity. The sequencing purpose, technology, DNA library, protocol, among others, influence directly the quantity of primer-adapter sequences in the reads or DNA sequences. In order to assemble a genome, adapter sequences must be efficiently removed. This process is defined as trimming, clipping, or filtering. Neglecting this process can easily result in suboptimal downstream analyses. There are many tools for trimming adapter sequences, such as Trimmomatic [2], AdapterRemoval [3], or AlienTrimmer [4]. Scarcely, adapter sequences are found in assembled genomes. Removing complete and partial adapter sequences, while leaving valid sequence data intact, is not trivial. The quality of trimming algorithms, non-optimized parameterization, ambiguity between adapter sequences and the target genomes, and the inclusion of partial adapter sequences are some of the reasons why the trimming process is highly complex [5].

The process of assembling a genome is also highly complex, primarily because it is dependent on specific technologies and techniques. Computationally, there may be multiple gaps, tiling path errors, and regions represented by uncommon alleles [6]. Other complex issues are inversions [7] and segmental duplications [8]. Also, distances with a constant size between exact subsequences of DNA have been found [9], where some were found to have biological meaning [10], while others remain unknown [11]. Therefore, the improvement of assemblies given the alignment of new sequencing reads remains a critical aspect of data interpretation [12], although continuous improvements of multiple assemblies are being attained.

Archaeal genomes are among those with the lowest number of genomes and associated studies, perhaps due to the extreme living environments and its associated difficulty to obtain the samples. Despite the size similarity to bacteria, archaea possess genes and several metabolic pathways that are more similar to those of eukaryotes, namely the enzymes involved, in transcription and translation, and their reliance on ether lipids in their cell membranes.

In this paper, we investigate the occurrence of exact or approximate primer and adapter subsequences in assembled genomes, specifically in the whole set of archaeal genomes from the NCBI database. We present a new method that combines data compression with custom signal processing operations, namely binary segmentation, filtering and segmentation, to localize and visualize these regions given a defined similarity threshold. The presentation of the method is given in the next section. In the results section, we use two controls to test the method, namely using pseudo-random sequences and substitution tests. Then, we present the localization maps of the whole archaeal genomes. Finally, we conclude and point to future directions.

2 Method

We propose a method composed of five phases, where a relative data compressor is combined in cascade with custom signal processing operations. The relative compressor uses a soft-blending mechanism, with a specific forgetting factor, between multiple context models and substitutional tolerant context models. The mentioned signal processing operations include binary segmentation, filtering, and segmentation. The output is automatically represented as an image at the respective scale. The methodology of the approach is depicted in Fig. 1.

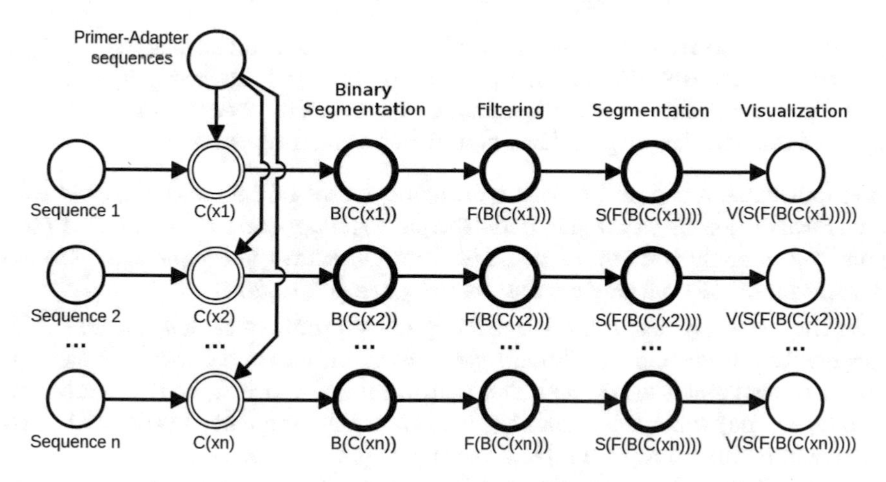

Fig. 1. Methodology for approximate localization of primer-adapter sequences. Function C stands for the relative compression of a certain sequence, x_i, given the primer-adapter sequences. Functions B, F, S and V correspond to Binary segmentation, Filtering, Segmentation, and Visualization, respectively.

As can be seen, we assume the presence of n DNA sequences. For example, in the case of the human genome, each sequence corresponds to a specific chromosome. The method compresses each sequence, x_i, using as a reference multiple primers or multiple adapter sequences, denoted by y. These sequences are well known from the wet and synthetic lab, namely because they are used in different scenarios and technologies (Illumina, ABI Solid, Ion Torrent, among others).

Since our purpose is to perform compression for analysis, instead of sending the probabilities of the mixture model to an arithmetic encoder, we compute a function that approximates with high precision the number of bits needed to compress each symbol, defined as

$$\mathcal{N}(x_{i,j}\|y) = -\log P(x_{i,j}\|y), \tag{1}$$

where $x_{i,j}$ represents the symbol of the sequence x_i and j is the position of a symbol along the sequence. The $P(x_{i,j}\|y)$ represents the predicted probability of the mixture model from the compressor relatively to y. The relative complexity profile is the sequence composed by each element of (1), assuming an increasing order, and denoted by $\overrightarrow{N}(x_i)$. Additional information on relative complexity profiles can be found in [13].

There are several referential compressors and some of them can be used for relative compression [14–18]. However, the majority are only efficient when a high similarity between the reference and the target sequence occur. The GeCo compressor [17] is efficiently prepared to deal with sequences that share higher divergence [19], mostly because it has a specific substitutional model [17,20]. Since our purpose is to find sequences that can have some degree of divergence [21], we use the GeCo compressor with the following four models:

(1) **tolerant context model**: depth: 18, alpha: 0.02, tolerance: 3;
(2) **context model**: depth: 18, alpha: 0.002, inverted repeats: yes;
(3) **context model**: depth: 13, alpha: 0.05, inverted repeats: yes;
(4) **context model**: depth: 11, alpha: 0.1, inverted repeats: yes.

The models are soft-blended with a decaying factor of 0.95 and a cache-hash of 3. It is worth noticing that the models expect the appearance of inverted repeat sequences, namely because of possible complementary subsequences. For more information on the parameters and models, see [17,19,20].

In the second phase, the relative complexity profile of each sequence, $\overrightarrow{N}(x_i)$, is segmented according to a binary segmentation, where a threshold, $t_B = 1.0$, is used to segment the regions. The real numbers above t_B are converted into the integer one, while the remaining into zero. The sequential order of the concatenation of the binary symbols gives the sequence $\mathcal{B}(x_{i,j})$.

In the third phase, the sequence $\mathcal{B}(x_{i,j})$ is filtered according to a low-pass filter, using a Blackman window of size 20, and sampled using a 1 out of 5 ratio. The output is a sequence of real numbers, $\mathcal{F}(x_{i,j})$, containing a smooth version of $\mathcal{B}(x_{i,j})$. In the fourth phase, the $\mathcal{F}(x_{i,j})$ is segmented according to another threshold where $t_S = 0.4$. The beginning and ending coordinates of the regions below t_S are recorded for further visualization, $\mathcal{S}(x_{i,j})$.

The last phase is the visualization of the coordinates of $\mathcal{S}(x_{i,j})$. For the purpose, we use a custom map built using Scalable Vector Graphics (SVG), where each region for each sequence is painted with a specific color.

We provide an implementation of the method, which for the whole set of tools uses less than 2 GB of RAM. The implementation code is freely available, under GPLv3 license, at https://github.com/pratas/maple.

3 Results

In this section, we apply the method on synthetic data to understand the sensitivity and fragmentation given increasing levels of substitutional mutations. Then, we apply the method to the full set of archaeal genomes from the NCBI (https://www.ncbi.nlm.nih.gov/). All the results of this paper can be fully replicated using an average laptop computer running a Linux operating system and the scripts provided at the repository: https://github.com/pratas/maple.

3.1 Synthetic Data Control

As a control process, we used synthetic genomic data. We created eight synthetic sequences $y_1, y_2, ..., y_8$, using XS [22] with a uniform distribution, with fixed sizes from 50 to 90 symbols. All these sequences, y, simulate custom primer-adapter sequences. Then, using a pseudo-random function to select one of the eight sequences, we made the sum of 1,000 copies and concatenated the sequences into one synthetic sample, x, of size 300,000 symbols, approximately. We made fourty copies of x and applied a specific increasing percentage of substitutional mutations over each x_i, where $i = \{1, 2, ..., 40\}$.

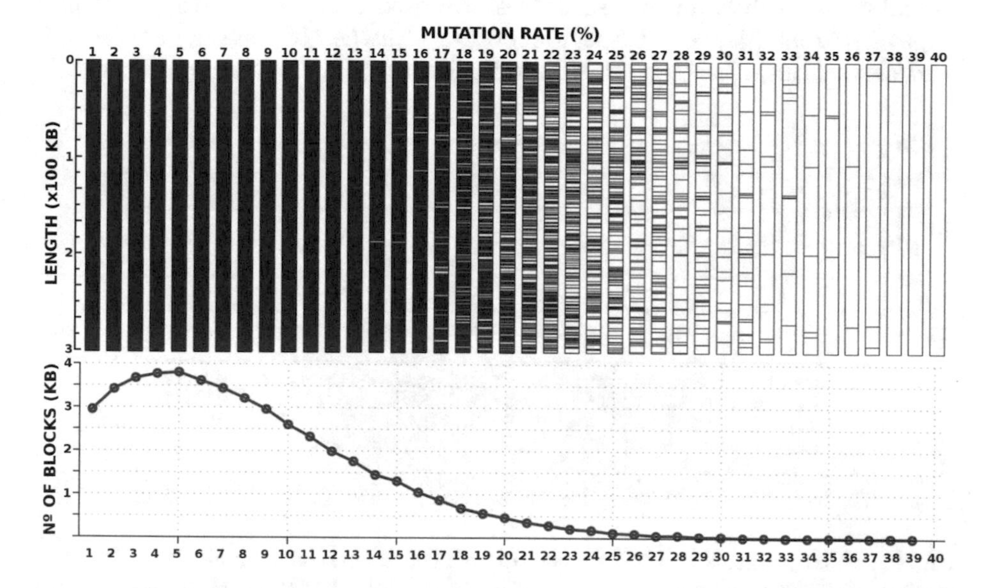

Fig. 2. Testing map (upper map) and the number of blocks (lower map) to an increasing percentage of substitutional mutations, over synthetic data, using the proposed method. The upper y-axis represents the length of the sequences (scale in 100 KB), while the lower y-axis the number of segmented blocks (scale in KB). The upper and lower x-axis are the same. The horizontal red stripes represent a region that the method found to be similar between y and each x_i. Each red line has been enlarged to a factor of 1,000 for visualization enhancement purposes. The experience can be fully replicated using the script `Synthetic.sh` from the code repository.

Figure 2 depicts the output map of the proposed method. As can be seen, the method can handle high levels of substitutions (high sensitivity). Since the method uses a lossless compression scheme, the results do not contain false positives. Each region contains a solid similarity and, hence, the method does not overestimate. For an extended demonstration of non-overestimation, see [23]. Regarding the number of segmented blocks, we see an increase in the first 5% of mutation, given the increase of fragmentation associated with the decrease in the length of the regions. Then, it decreases because, although the fragmentation increases, the number of similar mapped regions decreases faster. With 39% of substitutions, the method does not find any region. This shows that every symbol has a probability ≈0.4 of an edition. Hence, the probability of five symbols not having any edition is ≈0.07. Therefore, it is very hard to have methods that are able to detect similarity beyond these mutation rates without overestimation.

3.2 Archaeal Assembled Genomes

Given the need to maintain the scale and effective positions of the regions, we substituted the "N" regions (unknown bases in the process of sequencing or assembly) by pseudo-random sequences. We used a list of 152 primer-adapter sequences from Stephen Turner's repository (https://raw.githubusercontent.com/stephenturner/adapters/master/adapters_combined_152_unique.fasta).

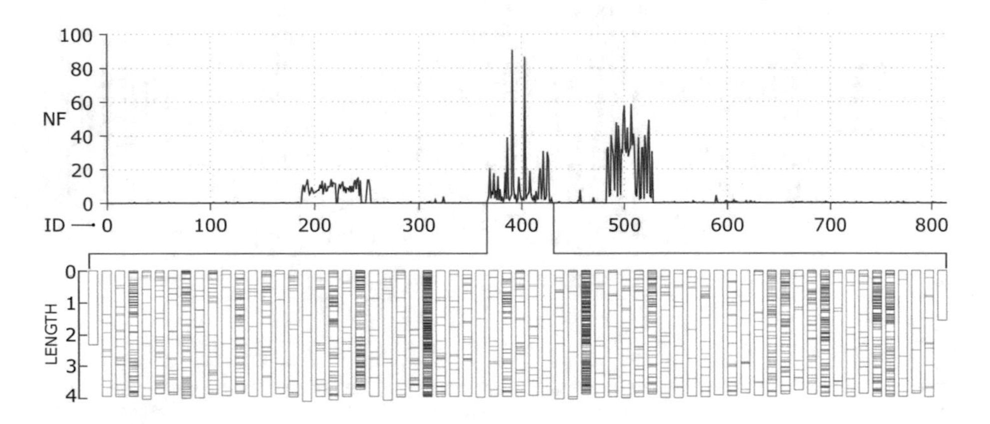

Fig. 3. Similar primer-adapter sequences in archaeal assembled genomes. The NF (Normalized quantity of Fragments) is computed using $F/|x_i| \times 100,000$, where F is the number of fragments and $|x_i|$ the length of x_i. The bottom map is a subregion of the profile (upper). The ID stands for a exclusive number which identifies each genome. Each red line has been enlarged to a factor of 10,000 for visualization purposes. The length is in Mega Bases. The experience can be fully replicated using the script `Maple_dir.sh` from the code repository.

Figure 3 depicts a high number of primer-adapter sequences in several archaeal assemblies of the NCBI database (7th February 2019). We are able to notice a high concentration of primer-adapter sequences in specific genomes, namely *Natronococcus jeotgali*, *Methanosarcina sp.*, and *Sulfolobus acidocaldarius*. The number of fragments relative to the size of the respective genomes times a constant (NF) shows that these three species have the highest number of similar fragments concentration.

We extended the work to identify the primer-adapter sequences with the highest similarity relative to the similar regions of these genomes. For the purpose, we extracted all the regions from the genome with flanking a margin of 20 bases. Then, we measured the similarity (using compression [24]) of each primer-adapter sequence to the three extracted sequences (mentioned previously). This measure can be replicated using script `Top_PA.sh`. The top similarity results, with the respective genomes, show the following adapters:

- *Natronococcus jeotgali* ($ID = 242$), I5 Nextera Transposase 2, `CTGTCTCTTATACACATCTCTGATGGCGCGAGGGAGGC`;
- *Methanosarcina sp.* ($ID = 403$), I5 Primer Nextera XT and Nextera Enrichment [N/S/E]504, `GACGCTGCCGACGATCTACTCTGTGTAGATCTCGGTGGTCGCCGTATCATT`;
- *Sulfolobus acidocaldarius* ($ID = 507$), Trans1, I5 Primer Nextera XT and Nextera Enrichment [N/S/E]508, I5 Primer Nextera XT Index Kit v2 S508 `TCGTCGGCAGCGTCAGATGTGTATAAGAGACAG` and `GACGCTGCCGACGAAGGCTTAGGTGTAGATCTCGGTGGTCGCCGTATCATT`.

4 Conclusions

We investigated the occurrence of exact and approximate primer-adapter subsequences in the whole set of the archaeal assembled genomes from the NCBI. We presented a new method that combines data compression with custom signal processing operations, namely filtering and segmentation, for localization and visualization of these regions given a defined similarity threshold. We have shown that the method is tolerant to high levels of substitutional mutations. We localized relative to primer-adapter sequences an unexpectedly high level of similar sequences, with distributed localization through all their genomes, in three species, namely *Natronococcus jeotgali*, *Methanosarcina sp.*, and *Sulfolobus acidocaldarius*. Further works include the analysis and comparison of assemblies from other domains.

Acknowledgments. This work was partially funded by FEDER (Programa Operacional Factores de Competitividade - COMPETE) and by National Funds through the FCT, in the context of the projects `UID/CEC/00127/2019` & `PTCD/EEI-SII/6608/2014` and the grant `PD/BD/113969/2015` to MH.

References

1. Pereira, F., Carneiro, J., Amorim, A.: Identification of species with DNA-based technology: current progress and challenges. Recent. Pat. DNA Gene Seq. **2**(3), 187–200 (2008)
2. Bolger, A.M., Lohse, M., Usadel, B.: Trimmomatic: a flexible trimmer for illumina sequence data. Bioinformatics **30**(15), 2114–2120 (2014)
3. Schubert, M., Lindgreen, S., Orlando, L.: AdapterRemoval v2: rapid adapter trimming, identification, and read merging. BMC Res. Notes **9**(1), 88 (2016)
4. Criscuolo, A., Brisse, S.: AlienTrimmer: a tool to quickly and accurately trim off multiple short contaminant sequences from high-throughput sequencing reads. Genomics **102**(5), 500–506 (2013)
5. Li, J.W., Bolser, D., Manske, M., Giorgi, F.M., Vyahhi, N., Usadel, B., Clavijo, B.J., Chan, T.F., Wong, N., Zerbino, D., et al.: The NGS WikiBook: a dynamic collaborative online training effort with long-term sustainability. Brief. Bioinform. **14**(5), 548–555 (2013)
6. Church, D., Deanna, M., Schneider, V., et al.: Modernizing reference genome assemblies. PLoS Biol. **9**(7), e1001091 (2011)

7. Hosseini, M., Pratas, D., Pinho, A.J.: On the role of inverted repeats in DNA sequence similarity. In: PACBB-2017, pp. 228–236 (2017)
8. Numanagić, I., Gökkaya, A.S., Zhang, L., Berger, B., Alkan, C., Hach, F.: Fast characterization of segmental duplications in genome assemblies. Bioinformatics **34**(17), i706–i714 (2018)
9. Afreixo, V., Bastos, C.A.C., Pinho, A.J., Garcia, S.P., Ferreira, P.J.S.G.: Genome analysis with inter-nucleotide distances. Bioinformatics **25**(23), 3064–3070 (2009)
10. Bastos, C.A., Afreixo, V., Rodrigues, J.M., Pinho, A.J.: An analysis of symmetric words in human DNA: adjacent vs non-adjacent word distances. In: PACBB-2018, pp. 80–87 (2018)
11. Tavares, A.H., Pinho, A.J., Silva, R.M., Rodrigues, J.M., Bastos, C.A., Ferreira, P.J., Afreixo, V.: DNA word analysis based on the distribution of the distances between symmetric words. Sci. Rep. **7**(1), 728 (2017)
12. Alkan, C., Sajjadian, S., Eichler, E.E.: Limitations of next-generation genome sequence assembly. Nat. Methods **8**(1), 61 (2010)
13. Pratas, D.: Compression and analysis of genomic data. Ph.D. thesis, University of Aveiro (2016)
14. Wandelt, S., Leser, U.: FRESCO: referential compression of highly similar sequences. IEEE/ACM Trans. Comput. Biol. Bioinform. **10**(5), 1275–1288 (2013)
15. Ochoa, I., Hernaez, M., Weissman, T.: iDoComp: a compression scheme for assembled genomes. Bioinformatics **31**, 626–633 (2014)
16. Deorowicz, S., Danek, A., Niemiec, M.: GDC 2: compression of large collections of genomes. Sci. Rep. **5**(11565), 1–12 (2015)
17. Pratas, D., Pinho, A.J., Ferreira, P.J.S.G.: Efficient compression of genomic sequences. In: DCC-2016, Snowbird, Utah, pp. 231–240 (2016)
18. Liu, Y., Peng, H., Wong, L., Li, J.: High-speed and high-ratio referential genome compression. Bioinformatics **33**(21), 3364–3372 (2017)
19. Pratas, D., Silva, R.M., Pinho, A.J.: Comparison of compression-based measures with application to the evolution of primate genomes. Entropy **20**(6), 393 (2018)
20. Pratas, D., Hosseini, M., Pinho, A.J.: Substitutional tolerant Markov models for relative compression of DNA sequences. In: PACBB-2017, pp. 265–272 (2017)
21. Crochemore, M., Ilie, L., Rytter, W.: Repetitions in strings: algorithms and combinatorics. Theor. Comput. Sci. **410**(50), 5227–5235 (2009)
22. Pratas, D., Pinho, A.J., Rodrigues, J.M.O.S.: XS: a FASTQ read simulator. BMC Res. Notes **7**(1), 40 (2014)
23. Pratas, D., Pinho, A.J., Silva, R.M., Rodrigues, J.M.O.S., Hosseini, M., Caetano, T., Ferreira, P.J.S.G.: FALCON-meta: a method to infer metagenomic composition of ancient DNA. bioRxiv 267179 (2018)
24. Garcia, S.P., Rodrigues, J.M.O.S., Santos, S., Pratas, D., Afreixo, V., Bastos, C.A.C., Ferreira, P.J.S.G., Pinho, A.J.: A genomic distance for assembly comparison based on compressed maximal exact matches. IEEE/ACM Trans. Comput. Biol. Bioinform. **10**(3), 793–798 (2013)

GeCo2: An Optimized Tool for Lossless Compression and Analysis of DNA Sequences

Diogo Pratas, Morteza Hosseini, and Armando J. Pinho(✉)

IEETA/DETI, University of Aveiro, Aveiro, Portugal
{pratas,seyedmorteza,ap}@ua.pt

Abstract. The development of efficient DNA data compression tools is fundamental for reducing the storage, given the increasing availability of DNA sequences. The importance is also reflected for analysis purposes, given the search for optimized and new tools for anthropological and biomedical applications. In this paper, we describe the characteristics and impact of the GeCo2 tool, an improved version of the GeCo tool. In the proposed tool, we enhanced the mixture of models, where each context model or tolerant context model has now a specific decay factor. Additionally, specific cache-hash sizes and the ability to run only a context model with inverted repeats was developed. A new command line interface, twelve new pre-computed levels, and several optimizations in the code were also included. The results show a compression improvement using less computational resources (RAM and processing time). This new version permits more flexibility for compression and analysis purposes, namely a higher ability of addressing different characteristics of the DNA sequences. The decompression is performed using symmetric computational resources (RAM and time). The GeCo2 is freely available, under GPLv3 license, at https://github.com/pratas/geco2.

Keywords: Data compression · Genomic sequence compression · GeCo2 tool · DNA sequences · Lossless data compression · Mixture models

1 Introduction

With the advent of the next-generation sequencing technologies associated to the decrease of sequencing costs [1], metagenomic studies [2], increasing availability of ancient genomes [3], and biomedical applications [4], the number of available complete genomic sequences increases dramatically, rendering the data deluge phenomenon a serious problem in most genomics centers [5]. General purpose compression algorithms are unable to attain the efficiency of specific purpose algorithms [6], therefore DNA sequence compressors are of great importance [7].

Genomic (or DNA) sequences are symbolic sequences, from an alphabet of four symbols $\Theta = \{A, C, G, T\}$, containing instructions, structure, and historic

F. Fdez-Riverola et al. (Eds.): PACBB 2019, AISC 1005, pp. 137–145, 2020.
https://doi.org/10.1007/978-3-030-23873-5_17

marks and trends of cellular organisms [8]. General purpose algorithms, such as gzip, bzip, 7zip, lzma, in the beginning were used to compress genomic sequences. Since the emergence of BioCompress [9], the development of specific genomic sequence compression algorithms revolutionized the field.

These sequences have multiple specific properties, namely high copy number, high level of substitutional mutations, high heterogeneity, and unknown level of noise, usually given to sequencing-assembly errors due to technology limitation and high ambiguity. To aggravate, there is contamination and the possibility of multiple rearrangements, such as inverted repeats and translocations. Despite the usage of general purpose algorithms with complex computational models, such as neural networks, to compress genomic sequences, the specific compressors, that efficiently address the specific nature of genomic sequences, show higher compression capabilities (5–10%) using substantially less resources [6].

The first specific algorithm, Biocompress, is based on a Lempel-Ziv dictionary approach [10], exploring repeats and palindromes. The Biocompress2 [11] is an extension of Biocompress [9], adding arithmetic coding of order-2 as a fallback mechanism. Since Biocompress, many algorithms have been proposed, namely Cfact [12], CDNA [13], ARM [14], Off-line [15], DNACompress [16], CTW + LZ [17], NMLComp [18], GeNML [19], DNA-X [20], DNAC [21], XM [22], 2D [23], DNASC [24], DBC [25], DNACompact [26], POMA [27], DNAEnc3 [28], DNAEnc4v2 [29], LUT [30], GenCodex [31], BIND [32], DNA-COMPACT [33], HighFCM [34], SeqCompress [35], HSFC [36], CoGI [37], GeCo [38], and OCW [39]. Also, reducing the storage size with combined encryption techniques was proposed [40,41]. For a comprehensive review, see [42].

Ranking the mentioned algorithms is a complex task. For example, some of these algorithms have been a contribution to other extensions or applications, while others are specialized for specific types of genomic sequences, such as bacteria, collections of genomes, and alignment data. There are also algorithms to cope with low computational resources. From our experience, we would highlight XM [22] and GeCo [38] given their ability to compress genomic sequences with high compression ratios. On average, XM is slightly better in relation to compression ratio (below 0.5% over GeCo). However, XM uses substantially higher RAM and time than GeCo [43]. Moreover, when the sequences to compress are larger than, say, 100 to 200 MB, the RAM increases to values that can not be supported by regular laptops, while GeCo uses consistently the same RAM, which can cope with the specific needs of regular laptops [43].

In this paper, we describe the characteristics and impact of the GeCo2 tool, which is an improved version of the GeCo tool. We enhanced the mixture of models, where each context model or substitutional tolerant context model has, now, a specific decay factor, allowing a better adaptation to the characteristics of each model depth and sequence nature. Additionally, specific cache-hash sizes and the ability to run exclusively a model with inverted repeats is developed. A new command line interface, twelve new pre-computed levels (optimized levels), and several optimizations in the code are also included.

In the remaining of the paper, we describe the improvements made in the new version. Then, we present the comparative results of the proposed compressor against state-of-the-art algorithms in a fair and consistent benchmark proposed in [43]. Finally, we draw some conclusions.

2 Method

GeCo [38] uses a soft-blending cooperation with a global forgetting factor between context models and substitutional tolerant context models [45], followed by arithmetic encoding. It has sub-programs to deal with inverted repeats and uses cache-hashes [46] for deeper models with a fixed equal size. The new version, GeCo2, instead of one forgetting factor, has a specific forgetting factor for each model. Accordingly, the probability of the next symbol, x_{n+1}, is given by

$$P(x_{n+1}) = \sum_{m \in \mathcal{M}} P_m(x_{n+1}|x_{n-k+1}^{n})\, w_{m,n}, \tag{1}$$

where $P_m(x_{n+1}|x_{n-m+1}^{n})$ is the probability assigned to the next symbol by a context [44] or substitutional tolerant context model [45], and where $w_{m,n}$ denote the corresponding weighting factors, with

$$w_{m,n} \propto (w_{m,n-1})^{\gamma_m} P_m(x_n|x_{n-k}^{n-1}), \quad \text{subjected to} \sum_{m \in \mathcal{M}} w_{m,n} = 1, \tag{2}$$

where $\gamma_m \in [0, 1)$ acts as a forgetting factor for each context model. Notice that the fundamental difference in the weights is given by setting each γ_m. Generally, we have found by exhaustive search that models with lower m are more prone to use lower γ_m (typically, below 0.9), while higher m, is associated with higher γ_m (near 0.95). This means that, in this mixture type, the forgetting intensity should be lower for more complex models. A curious indication was also found for a context model of order six. This model seems to be efficient with $\gamma_m \in [0.75; 0.85] \rightarrow k = 6$, which is the lowest γ_m among the models.

Besides the cooperation improvement, we added several functionalities depicted below. In the previous version, each model higher than a context order of 14 uses a hash table where it keeps only the latest *ch* hash collisions. In the current version, each model with context order higher than 14 has its own cache size ch_m. This enables to explore different repetitive natures without collision of models. In the previous version, the inverted repeats (IR) model could only be added to an existing model. In the current version, we added a mode to allow to run exclusively with inverted repeats. This allows to use the GeCo2 in inverted repeats studies, namely for the detection of rearrangements of inverted nature. We created a new interface layout, which enables more flexibility and permits easier optimization of parameters. New functions were added, namely a new approximate power function, and others optimized, having direct impact on decreasing the running time. We created 12 new pre-computed modes for reference-free compression, requiring only the flag "-l <*level*>" to be specified. The modes with the respective parameters are depicted in Table 1.

Table 1. Description of the parameters for the new 12 modes of GeCo2. For more information on the parameters meaning, see [38,45,46].

Mode	Context	Alpha	IR	Cache	Gamma	Tolerance	Alpha	Gamma
1	1	1	No	-	0.7	No	-	-
	12	1/20	Yes	-	0.97	No	-	-
2	2	1	No	-	0.78	No	-	-
	4	1	Yes	-	0.78	No	-	-
	11	1/80	Yes	-	0.96	No	-	-
3	3	1	No	-	0.80	No	-	-
	4	1	Yes	-	0.84	No	-	-
	12	1/50	Yes	-	0.94	2	1/15	0.95
4	4	1	No	-	0.80	No	-	-
	6	1	Yes	-	0.84	No	-	-
	13	1/50	Yes	-	0.94	2	1/15	0.95
5	4	1	No	-	0.82	No	-	-
	6	1	Yes	-	0.72	No	-	-
	13	1/50	Yes	-	0.95	2	1/15	0.95
6	4	1	No	-	0.88	No	-	-
	6	1	Yes	-	0.76	No	-	-
	13	1/50	Yes	-	0.95	2	1/15	0.95
7	4	1	Yes	-	0.90	No	-	-
	6	1	Yes	-	0.79	No	-	-
	8	1	Yes	-	0.91	No	-	-
	13	1/10	Yes	-	0.94	1	1/20	0.94
	16	1/200	Yes	5	0.95	4	1/15	0.95
8	4	1	Yes	-	0.90	No	-	-
	6	1	Yes	-	0.80	No	-	-
	13	1/10	Yes	-	0.95	1	1/20	0.94
	16	1/100	Yes	5	0.95	3	1/15	0.95
9	4	1	Yes	-	0.91	No	-	-
	6	1	Yes	-	0.82	No	-	-
	13	1/10	Yes	-	0.94	1	1/20	0.94
	17	1/100	Yes	8	0.95	3	1/15	0.95
10	1	1	No	-	0.90	No	-	-
	3	1	No	-	0.90	No	-	-
	6	1	Yes	-	0.82	No	-	-
	9	1/10	No	-	0.90	No	-	-
	11	1/10	No	-	0.90	No	-	-
	13	1/10	Yes	-	0.90	No	-	-
	17	1/100	Yes	8	0.89	5	1/10	0.90
11	4	1	Yes	-	0.91	No	-	-
	6	1	Yes	-	0.82	No	-	-
	13	1/10	Yes	-	0.95	1	1/20	0.94
	17	1/100	Yes	15	0.95	3	1/15	0.95
12	1	1	No	-	0.90	No	-	-
	3	1	No	-	0.90	No	-	-
	6	1	Yes	-	0.85	No	-	-
	9	1/10	No	-	0.90	No	-	-
	11	1/10	No	-	0.90	No	-	-
	13	1/50	Yes	-	0.90	No	-	-
	17	1/100	Yes	20	0.90	3	1/10	0.90

3 Results

We benchmark GeCo2 with state-of-the-art compression tools using a fair dataset proposed in [43]. This dataset contains 534,263,017 bases (509.5 MB) with 15 DNA sequences of several sizes. It includes sequences from different domains and kingdoms, namely viruses, archaea, bacteria and eukaryota.

All the results presented in this paper were computed in a Linux server with a single core Intel Xeon CPU E7320 at 2.13 GHz. The tool (GeCo2) has been implemented in C language and is available, under the GPLv3 license, at http://github.com/pratas/geco2. We ran paq8 with the -8 (best option), GeCo using "-tm 1:1:0:0/0 -tm 3:1:0:0/0 -tm 6:1:0:0/0 -tm 9:10:0:0/0 -tm 11:10:0:0/0 -tm 13:50:1:0/0 -tm 18:100:1:3/10 -c 30 -g 0.9" and XM using 50 copy experts.

We present the comparative results of the proposed compressor (GeCo2) against state-of-the-art algorithms, namely paq8, GeCo and XM. Table 2 depicts the number of bytes needed to compress each DNA sequence for each compressor and Table 3 the respective computational time.

Table 2. Number of bytes needed to represent each DNA sequence (individually) given the respective data compressor (paq8 -8, GeCo, XM and GeCo2). The compression level (mode) used in GeCo2 is within parenthesis.

ID	paq8 -8	GeCo	XM	GeCo2 (mode)
HoSa	40,517,624	38,877,294	38,940,458	**38,845,642** (12)
GaGa	34,490,967	33,925,250	33,879,211	**33,877,671** (11)
DaRe	12,628,104	11,520,064	**11,302,620**	11,488,819 (10)
OrSa	9,280,037	8,671,732	**8,470,212**	8,646,543 (10)
DrMe	7,577,068	7,498,808	7,538,662	**7,481,093** (10)
EnIn	5,761,090	5,196,083	**5,150,309**	5,170,889 (9)
ScPo	2,557,988	2,536,457	2,524,147	**2,518,963** (8)
PlFa	1,959,623	1,944,036	1,925,841	**1,925,726** (7)
EsCo	1,107,929	1,109,823	1,110,092	**1,098,552** (6)
HaHi	904,074	906,991	913,346	**902,831** (5)
AeCa	380,273	385,640	387,030	**380,115** (5)
HePy	385,096	381,545	384,071	**375,481** (4)
YeMi	16,835	17,167	16,861	**16,798** (3)
AgPh	10,754	10,882	10,711	**10,708** (2)
BuEb	4,668	4,774	**4,642**	4,686 (1)
Total	117,582,130	112,986,546	**112,558,213**	112,744,517

Compared with GeCo, GeCo2 is able to compress better 0.2142% the complete dataset (using $100 - \{C_{other}/C_{best} \times 100\}$), while using less 11.8% of computational time. The GeCo (version 1) uses ~4.8 GB of RAM. This version,

Table 3. Computational time (in seconds) needed to compress (individually) the dataset for each data compressor (paq8, GeCo, XM and GeCo2). The compression level (mode) used in GeCo2 is according to Table 2.

ID	paq8	GeCo	XM	GeCo2
HoSa	85,269.1	**648.6**	5,589.8	652.4
GaGa	64,898.9	503.2	3,633.9	**494.7**
DaRe	29,907.7	215.9	785.2	**198.8**
OrSa	20,745.1	192.4	489.7	**138.3**
DrMe	14,665.8	114.6	362.6	**102.4**
EnIn	11,183.6	95.8	279.8	**82.5**
ScPo	4,619.1	45.2	96.5	**34.2**
PlFa	4,133.9	39.7	84.4	**35.3**
EsCo	1,973.9	26.4	36.8	**5.1**
HaHi	1,738.1	23.7	39.1	**4.4**
AeCa	675.3	17.0	10.3	**1.9**
HePy	715.1	17.2	11.2	**1.9**
YeMi	32.6	12.3	0.9	**0.1**
AgPh	20.1	12.1	0.9	**0.1**
BuEb	9.1	12.2	0.7	**0.1**
Total	240,587.4	1,976.3	11,421.8	**1,742.2**

GeCo2, used at most ∼3.8 GB of RAM. Therefore, besides the improvement in the compression while using less computational time, we saved ∼1 GB of RAM.

Compared with GeCo2, the XM method was able to compress significantly better the OrSa and DaRe sequences. If we exclude these sequences from the dataset, in the remaining sequences, GeCo2 would compress better than XM. Therefore, the difference is in the nature of these sequences. The OrSa and DaRe sequences represent rice and fish genomes, respectively. Future works will focus on the development of efficient models for these sequences. Nevertheless, XM was able to compress 0.1652% better than GeCo2 but using 6.5× more the computational time and substantially higher RAM (at least 3×) than GeCo2. Notice GeCo, GeCo2, and XM decompressions are approximately symmetric since they compute the corresponding decompression using approximately equal computational resources (time and RAM).

Regarding general purpose data compressors, GeCo2 was able to compress the dataset 4, 1142% better than paq8 using 138× less the computational time (paq8 RAM did not exceed 2 GB). This difference shows that, unlike the best general purpose algorithms (in the best mode), GeCo2 is efficient.

4 Conclusions

The development of efficient data compression algorithms is fundamental for reducing the storage of sequencing projects. The importance is also at the analysis level, with direct implications in the anthropological and biomedical field.

In this paper, we described the characteristics and impact of GeCo2, the improved version of GeCo. We enhanced the mixture of the models, added specific cache-hash sizes, exclusive use of inverted repeats, a new command line interface, twelve new pre-computed levels, and several code optimizations.

The results show a compression improvement using less computational resources. Moreover, this version allows a higher flexibility for analysis purposes.

Acknowledgments. This work was partially funded by FEDER (Programa Operacional Factores de Competitividade - COMPETE) and by National Funds through the FCT, in the context of the projects UID/CEC/00127/2019 & PTCD/EEI-SII/6608/2014 and the grant PD/BD/113969/2015 to MH.

References

1. Mardis, E.R.: DNA sequencing technologies: 2006–2016. Nat. Protoc. **12**(2), 213 (2017)
2. Marco, D.: Metagenomics: Theory, Methods and Applications. Horizon Scientific Press, Poole (2010)
3. Marciniak, S., et al.: Harnessing ancient genomes to study the history of human adaptation. Nat. Rev. Genet. **18**(11), 659 (2017)
4. Weber, W., et al.: Emerging biomedical applications of synthetic biology. Nat. Rev. Genet. **13**(1), 21 (2012)
5. Schatz, M.C., et al.: The DNA data deluge. IEEE Spectr. **50**(7), 28–33 (2013)
6. Goyal, M., et al.: DeepZip: lossless data compression using recurrent neural networks. arXiv:1811.08162 (2018)
7. Sayood, K.: Introduction to Data Compression. Morgan Kaufmann, Burlington (2017)
8. Dougherty, E.R., et al. (eds.): Genomic Signal Processing and Statistics. Hindawi Publishing Corporation, London (2005)
9. Grumbach, S., et al.: Compression of DNA sequences. In: DCC-1993, Utah, pp. 340–350 (1993)
10. Ziv, J., et al.: A universal algorithm for sequential data compression. IEEE Trans. Inf. Theory **23**, 337–343 (1977)
11. Grumbach, S., et al.: A new challenge for compression algorithms: genetic sequences. Inf. Process. Manag. **30**(6), 875–886 (1994)
12. Rivals, E., et al.: A guaranteed compression scheme for repetitive DNA sequences. In: DCC-1996, Utah, p. 453 (1996)
13. Loewenstern, D., et al.: Significantly lower entropy estimates for natural DNA sequences. In: DCC-1997, Utah (1997)
14. Allison, L., et al.: Compression of strings with approximate repeats. In: Proceedings of Intelligent Systems in Molecular Biology, ISMB 1998, Montreal, Canada, pp. 8–16 (1998)

15. Apostolico, A., et al.: Compression of biological sequences by greedy off-line textual substitution. In: DCC-2000, Utah (2000)
16. Chen, X., et al.: DNACompress: fast and effective DNA sequence compression. Bioinformatics **18**(12), 1696–1698 (2002)
17. Matsumoto, T., et al.: Biological sequence compression algorithms. In: Proceedings of the 11th Workshop, Tokyo, Japan, pp. 43–52 (2000)
18. Tabus, I., et al.: DNA sequence compression using the normalized maximum likelihood model for discrete regression. In: DCC-2003, Utah, pp. 253–262 (2003)
19. Korodi, G., et al.: An efficient normalized maximum likelihood algorithm for DNA sequence compression. ACM Trans. Inf. Syst. **23**(1), 3–34 (2005)
20. Manzini, G., et al.: A simple and fast DNA compressor. Softw.—Pract. Exper. **34**, 1397–1411 (2004)
21. Lee, A.J.T., et al.: DNAC: an efficient compression algorithm for DNA sequences. National Taiwan University, Taipei 10617, R.O.C. 1(1) (2004)
22. Cao, M.D., et al.: A simple statistical algorithm for biological sequence compression. In: DCC-2007, Utah (2007)
23. Vey, G.: Differential direct coding: a compression algorithm for nucleotide sequence data. Database (2009)
24. Mishra, K.N., et al.: An efficient horizontal and vertical method for online DNA sequence compression. Int. J. Comput. Appl. **3**(1), 39–46 (2010)
25. Rajeswari, P.R., et al.: GENBIT Compress-Algorithm for repetitive and non repetitive DNA sequences. Int. J. Comput. Sci. Inf. Technol. **2**, 25–29 (2010)
26. Gupta, A., et al.: A novel approach for compressing DNA sequences using semi-statistical compressor. Int. J. Comput. Appl. **33**, 245–251 (2011)
27. Zhu, Z., et al.: DNA sequence compression using adaptive particle swarm optimization-based memetic algorithm. IEEE Trans. Evol. Comput. **15**(5), 643–658 (2011)
28. Pinho, A.J., et al.: Bacteria DNA sequence compression using a mixture of finite-context models. In: IEEE Workshop on Statistical Signal Processing, Nice (2011)
29. Pinho, A.J., et al.: On the representability of complete genomes by multiple competing finite-context (Markov) models. PLoS ONE **6**(6), e21588 (2011)
30. Roy, S., et al.: An efficient biological sequence compression technique using LUT and repeat in the sequence. arXiv:1209.5905 (2012)
31. Satyanvesh, D., et al.: GenCodex - a novel algorithm for compressing DNA sequences on multi-cores and GPUs. In: Proceedings of IEEE 19th International Conference on High Performance Computing (HiPC), Pune (2012)
32. Bose, T., et al.: BIND-an algorithm for loss-less compression of nucleotide sequence data. J. Biosci. **37**(4), 785–789 (2012)
33. Li, P., et al.: DNA-COMPACT: DNA compression based on a pattern-aware contextual modeling technique. PLoS ONE **8**(11), e80377 (2013)
34. Pratas, D., et al.: Exploring deep Markov models in genomic data compression using sequence pre-analysis. In: EUSIPCO-2014, Lisbon, pp. 2395–2399 (2014)
35. Sardaraz, M., et al.: SeqCompress: an algorithm for biological sequence compression. Genomics **104**(4), 225–228 (2014)
36. Guo, H., et al.: Genome compression based on Hilbert space filling curve. In: International Conference on Management, Education, Information and Control (MEICI 2015), Shenyang, pp. 29–31 (2015)
37. Xie, X., et al.: CoGI: towards compressing genomes as an image. IEEE/ACM Trans. Comput. Biol. Bioinform. **12**(6), 1275–1285 (2015)
38. Pratas, D., et al.: Efficient compression of genomic sequences. In: DCC-2016, Utah, pp. 231–240 (2016)

39. Chen, M., et al.: Genome sequence compression based on optimized context weighting. Genet. Mol. Res.: GMR **16**(2) (2017)
40. Pratas, D., et al.: Cryfa: a tool to compact and encrypt FASTA files. In: PACBB-2017, pp. 305–312 (2017)
41. Hosseini, M., et al.: Cryfa: a secure encryption tool for genomic data. Bioinformatics **35**(1), 146–148 (2018)
42. Hosseini, M., et al.: A survey on data compression methods for biological sequences. Information **7**(4), 56 (2016)
43. Pratas, D., et al.: A DNA sequence corpus for compression benchmark. In: PACBB-2018, pp. 208–215 (2018)
44. Bell, T.C., et al.: Text Compression. Prentice Hall, Upper Saddle River (1990)
45. Pratas, D., et al.: Substitutional tolerant Markov models for relative compression of DNA sequences. In: PACBB-2017, pp. 265–272 (2017)
46. Ferreira, P.J.S.G., et al.: Compression-based normal similarity measures for DNA sequences. In: ICASSP-2014, Florence, pp. 419–423 (2014)

Troppo - A Python Framework for the Reconstruction of Context-Specific Metabolic Models

Jorge Ferreira(✉), Vítor Vieira, Jorge Gomes, Sara Correia, and Miguel Rocha

Centre of Biological Engineering, Campus de Gualtar, University of Minho, Braga, Portugal
{jorge.ferreira,vvieira}@ceb.uminho.pt, jorge.gomes12@gmail.com, sarag.correia@gmail.com, mrocha@di.uminho.pt

Abstract. The surge in high-throughput technology availability for molecular biology has enabled the development of powerful predictive tools for use in many applications, including (but not limited to) the diagnosis and treatment of human diseases such as cancer. Genome-scale metabolic models have shown some promise in clearing a path towards precise and personalized medicine, although some challenges still persist. The integration of omics data and subsequent creation of context-specific models for specific cells/tissues still poses a significant hurdle, and most current tools for this purpose have been implemented using proprietary software. Here, we present a new software tool developed in Python, *troppo* - Tissue-specific RecOnstruction and Phenotype Prediction using Omics data, implementing a large variety of context-specific reconstruction algorithms. Our framework and workflow are modular, which facilitates the development of newer algorithms or omics data sources.

Keywords: Context-specific model reconstruction · Tissue specific models · Genome-scale metabolic models · Omics data integration

1 Introduction

Over the past years, the relationship between biology and informatics has proven to work to unveil the mysteries of the cell, from sequencing genomes to the reconstruction of the metabolism for a human cell. With ongoing advances on high-throughput technologies, the scientific community has been able to widen its scope of research, being able to analyze the cell as a complex layered system of interactions [5].

Genome-scale metabolic models (GSMMs) are the result of the integration of genome information into Constraint-Based Models (CBMs), connecting the

F. Fdez-Riverola et al. (Eds.): PACBB 2019, AISC 1005, pp. 146–153, 2020.
https://doi.org/10.1007/978-3-030-23873-5_18

genotype with cell metabolism. GSMMs are normally used to predict phenotypes which can be associated with consumption/ production rates of one or several metabolites [14].

The first reconstructions to emerge, through extensive manual curation, were Recon1 [8] and Edinburgh Human Metabolic Network (EHMN) [13]. To date, the most recent and complete reconstruction is the third version of Recon, Recon3D, integrating most of all the previous reconstructed models and including information related to the human microbiome and metabolism of dietary compounds [6]. However, these models represent the metabolism of a generalized human cell, eliciting the need for tissue specific model reconstruction algorithms. One of the main advantages of GSMMs is the ability to easily integrate omics data, i.e. to generate tissue specific metabolic models, with several algorithms already developed by the scientific community in the last couple of decades [5,16].

These allow the tuning of generic models to specific tissues or cell types, allowing to perform context-specific phenotype predictions and analyses. Although there are diverse algorithms proposed towards this aim, these are implemented in different platforms. Some are available in COBRA toolbox, which is dependent on a commercial platform (MatLab) [10].

In this work, we aim to provide a novel software platform, in open-source software, developed in Python, *Troppo - Tissue-specific RecOnstruction and Phenotype Prediction using Omics data*, implementing a wide range of context-specific algorithms, that can take as input a generic model and different types of omics data to provide context-specific models and/ or phenotype predictions (flux distributions).

2 Methods

The evolution of omics technologies led to the birth of the algorithms for reconstruction of tissue specific metabolic models since 2010, with the development of the MBA algorithm [12]. Although these algorithms have a similar goal, they can be divided into three different groups according to some of its characteristics: the definition of a core (Model-Building Algorithm (MBA)-like algorithms), testing Required Metabolic Functions (RMF) (Gene Inactivation Moderated by Metabolism and Expression (GIMME)-like algorithms) and creation of threshold based on the gene/protein expression (integrative Metabolic Analysis Tool (iMAT)-like family) [7,15].

2.1 GIMME-Like Family

This family contains the original GIMME algorithm [3], GIMMEp (allowing the incorporation of proteomics data) [4] and its extension GIM^{3}E [17]. The basis of this family is to perform a reconstruction by first optimizing the objective function (RMF) through a Linear Programming (LP) formulation. A second LP minimizing a penalty function (related to differences of the values obtained from the flux analysis to the respective transcript ones) is also performed, adding a constraint requiring a RMF value above a certain lower bound.

These algorithms differ in the way the penalty function is defined. GIMME penalizes flux values of reactions which are below a defined threshold when compared to the associated expression values. As for the GIM^{3}E, the penalties are calculated for each reaction associated with expression levels (instead of a set in GIMME), allowing all the reactions to have a penalty score. Also, since GIM^{3}E includes thermodynamics constraints, it is formulated as a Mixed Integer Linear Programming (MILP) problem.

2.2 iMAT-Like Family

In this family, we find the algorithms iMAT [23], INIT [1] and tINIT [2], which also reconstruct a model based on experimental data. The main difference regarding the previous family is that they do not need a RMF to work, so they try to match the maximum possible number of reaction states (active) to related data expression. Their formulation is a MILP and despite they share a similar strategy, iMAT tries to incorporate data in the constraints of the model, while INIT and tINIT try do so on the objective function.

iMAT is based on a previously defined threshold for the expression data which separates the reactions in high and low expression. Next, after defining a minimum flux value for each group, it tries to match the maximum number of reactions to each group, and specifically for the highly expressed group, so it has to overcome the minimum flux by solving a MILP. However, due to the fact that there are several possible flux distributions which have the same objective function value, iMAT uses an adapted version of Flux Variability Analysis [23].

The INIT algorithm maximizes the matches between reaction states (active or inactive) and data regarding expression of genes/proteins, returning flux values and a tissue-specific model. The method solves a MILP, where binary variables represent the presence of each reaction from the template model in the resulting model. In the definition of the objective function, positive weights are given to reactions with a higher evidence from the input, and negative to the ones which have low or no expression. If there is supportive information (usually metabolomics) that corroborate the presence of a certain metabolite, the necessary reactions may be included in the final model to produce [1].

The tINIT is an extension of the previous algorithm [2]. The improvement is based on the possibility to define a set of metabolic tasks in agreement with the context of the reconstruction. These may be the consumption or production of a metabolite or activation of the reactions of a particular pathway for the tissue.

2.3 MBA-Like Family

The last family is constituted by MBA [12], mCADRE [21] and FASTCORE [20]. Unlike the previously described algorithms, these only return the resulting context-specific model. Their inputs are sets with predefined categorized reactions as core and non-core sets. Normally, the core is defined by the reactions which have higher evidence to be considered active (high-throughput data or curated biochemical information). When the core is built, the methods try to eliminate or include the non-core reactions, while ensuring that the model is consistent and has no blocked reactions when reconstructed.

For MBA, both cores are created according to the provided data. In an iterative way, all the non-core reactions are removed in a random order, while the model is tested for consistency. The iteration ends when all the reactions have been submitted for the removal test in the final model. Since the order by which reactions are tested for removal matters, there is the need to repeat this algorithm several times to obtain a set of models. The final one should be a model based on the ranking of the frequency of the reactions in the set, adding them to the high evidence core until a coherent model is found [12].

The mCADRE algorithm is quite similar to the MBA, but only requires the reconstruction of a single model. It is initialized by ranking the reactions on the original model using three distinct scores: confidence, expression and connectivity. With the help of a threshold value for the scores, a core of reactions and the order of removal of the non-core ones is established [21].

FASTCORE algorithm uses another strategy by solving two LPs. The first maximizes the number of reactions in the core, comparing the values of a reaction with a constant, while the other decreases the number of reactions that are absent in the core by minimizing the L_1-norm of the flux vector. Until the core is coherent (the whole set of core reactions is activated with the smallest number of non-core reactions), both problems are solved alternatively and in a repeated way. For reversible reactions, the algorithm analyses both directions [20].

CORDA is a relatively new algorithm and due to its nature for the predefinition of a core of reactions, it might be classified MBA-like algorithm. One of its main features is the fact of only needing a LP, providing a faster reconstruction in comparison to other algorithms. As a novel approach, the developers created the *dependency assessment* as a new way to identify the importance of desirable reactions (with higher evidence) in contrast to the one with less information [18].

An overview of all the algorithm families and how to choose the better algorithm for a certain situation is depicted in the Fig. 1.

3 Software

We have developed *troppo*, a modular framework implemented using the Python programming language and providing routines for the reconstruction of context-specific metabolic models (Fig. 1). *troppo* does not natively depend or provide any model reading or manipulation capabilities. This is intended, since external wrappers for cobrapy [9] and framed (https://github.com/cdanielmachado/framed) are available, allowing the user to load and manipulate previously validated models with these tools, and use them as inputs for the reconstruction algorithms provided in *troppo*. Additionally, the framework is built such that these operations are performed with short and simple commands.

Troppo depends on the cobamp library, an open-source tool implementing constraint-based pathway analysis methods, as the underlying framework

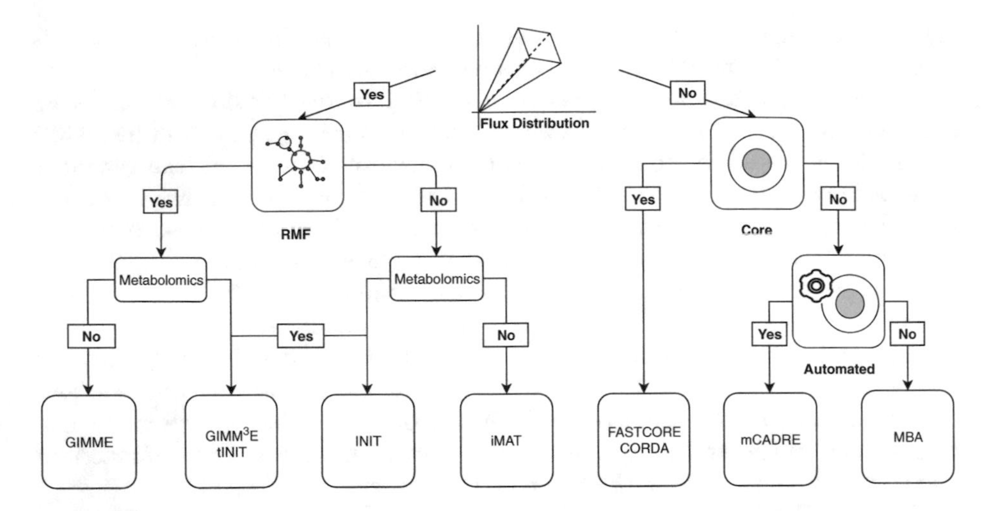

Fig. 1. Overview of the developed algorithms for context-specific model reconstruction and their properties, including the ability to obtain a flux distribution, as well as possible inputs such as metabolic functions (RMFs), metabolomics data or a core of curated reactions

providing methods to connect with external model loading frameworks and for building and solving LP/MILP problems. The latter is also dependent on *optlang* [11], allowing a wide selection of commercial and non-commercial solvers to be used.

The software is comprised of two main components, namely an omics data processing layer capable of handling transcriptomics and proteomics data, and a modular reconstruction layer implementing context-specific reconstruction methods. The general workflow implemented in this software is depicted on Fig. 2.

3.1 Omics Layer

Omics data is obtained using one of the provided readers for proteomics data from HPA [19] and as microarray experiments from any source. Using mappings from the HUGO Gene Nomenclature Committee [22], *troppo* is capable of mapping genes encoding enzymes with their respective reactions. This integration relies on gene-protein-reaction rules (GPRs) usually present on GSMMs, which can be used to modulate gene expression and adjust flux predictions accordingly. The preprocessed and mapped data is then passed as an algorithm property in the form of either weights for each reaction or sets of reactions, depending on the algorithm.

3.2 Reconstruction Layer

In the reconstruction layer, two key inputs are needed for any of the implemented algorithms. A metabolic model must be supplied **(2a)**, which at a bare minimum, implies a stoichiometric matrix encoding the relations between metabolites and reactions, as well as thermodynamic constraints for the latter (lower and upper bounds). All algorithms also require a matching set properties, individual to each method, where specific execution options and omics inputs pre-processed using the omics layer described in the previous section **(1)** are included **(2b)**.

After defining these inputs, the selected algorithm will be run **(3)** and the expected output is a set of reactions to exclude from the generic model **(4a)**. Some methods are also capable of returning a flux distribution constrained by omics data **(4b)**, which can later be used for analysis or to manually extract the inactive reactions from it. Finally, these reactions are excluded from the model, alongside with their respective genes **(5)**, completing the reconstruction process.

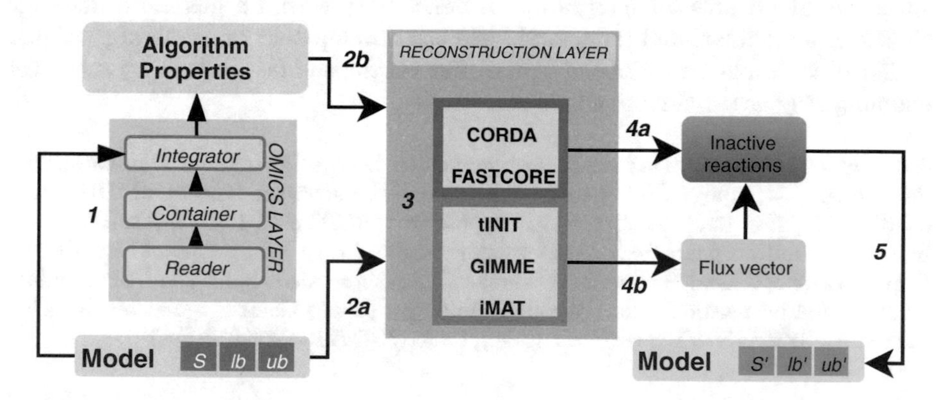

Fig. 2. Overview of the generic pipeline implemented within *troppo*. The process starts with a generic metabolic model - stoichiometric matrix (**S**), lower and upper bounds (**lb**, **ub**) - with pre-processed omics data associated with its identifiers **(1)**. This model as well as specific algorithm parameters, including those obtained using omics data, are the inputs for the reconstruction algorithms **(2)**, whose result is ultimately a set of inactive reactions **(3, 4)**. This final set is then removed from the model **(5)**

3.3 Availability

Troppo is available in the Python programming language, preferably from version 3.6 onwards. It is licensed under the GNU Public License (version 3.0), with source-code available on GitHub (https://github.com/BioSystemsUM/troppo). A LP/MILP solver is required to run the algorithms and our framework is currently compatible with CPLEX, Gurobi and GNU Linear Programming Kit (GLPK).

4 Results

The software featured in this work was used to reconstruct tissue-specific metabolic models of tumors arising from glial cells (glioma). To validate the implementations, we also used similar MATLAB routines based on the COBRA

Toolbox to run the algorithms which were implemented in *troppo* package using CPLEX as the undelying LP solver. These are then compared with solutions from *troppo*. These analyses are included in the GitHub repository as Jupyter notebooks which users can download and run (https://github.com/BioSystemsUM/troppo/tests/validation/).

5 Conclusion

In this work, a Python open-source library for context-specific model reconstruction was presented. The software currently integrates a large and growing variety of algorithms for integration of omics data within a generic framework that is also modular, and thus, facilitates the development of newer algorithms. Unlike other implementations, this one does not depend on proprietary software, enabling easy access for the whole community.

Acknowledgements. This study was supported by the Portuguese Foundation for Science and Technology (FCT) under the scope of the strategic funding of UID/BIO/04469/2019 unit and BioTecNorte operation (NORTE-01-0145-FEDER-000004) funded by the European Regional Development Fund under the scope of Norte2020 - Programa Operacional Regional do Norte. The authors also thank the PhD scholarships funded by national funds through Fundação para a Ciência e Tecnologia, with references: SFRH/BD/133248/2017 (J.F.), SFRH/BD/118657/2016 (V.V.).

References

1. Agren, R., Bordel, S., Mardinoglu, A., et al.: Reconstruction of genome-scale active metabolic networks for 69 human cell types and 16 cancer types using INIT. PLoS Comput. Biol. **8**(5), e1002518 (2012). https://doi.org/10.1371/journal.pcbi.1002518
2. Agren, R., Mardinoglu, A., Asplund, A., et al.: Identification of anticancer drugs for hepatocellular carcinoma through personalized genome-scale metabolic modeling. Mol. Syst. Biol. (2014). https://doi.org/10.1002/msb.145122
3. Becker, S.A., Palsson, B.O.: Context-specific metabolic networks are consistent with experiments. PLoS Comput. Biol. **4**(5), e1000082 (2008). https://doi.org/10.1371/journal.pcbi.1000082
4. Bordbar, A., Mo, M.L., Nakayasu, E.S., et al.: Model-driven multi-omic data analysis elucidates metabolic immunomodulators of macrophage activation. Mol. Syst. Biol. (2012). https://doi.org/10.1038/msb.2012.21
5. Bordbar, A., Monk, J.M., King, Z.A., Palsson, B.O.: Constraint-based models predict metabolic and associated cellular functions (2014). https://doi.org/10.1038/nrg3643
6. Brunk, E., Sahoo, S., Zielinski, D.C., et al.: Recon3D enables a three-dimensional view of gene variation in human metabolism. Nat. Biotechnol. **36**(3), 272–281 (2018). https://doi.org/10.1038/nbt.4072
7. Correia, S., Costa, B., Rocha, M.: Reconstruction of consensus tissue-specific metabolic models. bioRxiv (2018). https://doi.org/10.1101/327262

8. Duarte, N.C., Becker, S.A., Jamshidi, N., et al.: Global reconstruction of the human metabolic network based on genomic and bibliomic data. Proc. Nat. Acad. Sci. USA **104**(6), 1777–82 (2007). https://doi.org/10.1073/pnas.0610772104
9. Ebrahim, A., Lerman, J.A., Palsson, B.O., Hyduke, D.R.: COBRApy: constraints-based reconstruction and analysis for Python. BMC Syst. Biol. (2013). https://doi.org/10.1186/1752-0509-7-74
10. Heirendt, L., Arreckx, S., Pfau, T., et al.: Creation and analysis of biochemical constraint-based models: the COBRA toolbox v3.0. Protocol Exchange (2017). https://doi.org/10.1038/protex.2011.234
11. Jensen, K., G.R. Cardoso, J., Sonnenschein, N.: Optlang: an algebraic modeling language for mathematical optimization. J. Open Sour. Softw. (2017). https://doi.org/10.21105/joss.00139
12. Jerby, L., Shlomi, T., Ruppin, E.: Computational reconstruction of tissue-specific metabolic models: application to human liver metabolism. Mol. Syst. Biol. **6**, 401 (2010). https://doi.org/10.1038/msb.2010.56
13. Ma, H., Sorokin, A., Mazein, A., et al.: The Edinburgh human metabolic network reconstruction and its functional analysis. Mol. Syst. Biol. **3**(1), 135 (2007). https://doi.org/10.1038/msb4100177
14. Raškevičius, V., Mikalayeva, V., Antanavičičtūė, I., et al.: Genome scale metabolic models as tools for drug design and personalized medicine. PLOS ONE **13**(1), e0190636 (2018). https://doi.org/10.1371/journal.pone.0190636
15. Robaina Estévez, S., Nikoloski, Z.: Generalized framework for context-specific metabolic model extraction methods. Front. Plant Sci. **5**, 491 (2014). https://doi.org/10.3389/fpls.2014.00491
16. Ryu, J.Y., Kim, H.U., Lee, S.Y.: Reconstruction of genome-scale human metabolic models using omics data. Integr. Biol. (2015). https://doi.org/10.1039/C5IB00002E
17. Schmidt, B.J., Ebrahim, A., Metz, T.O., et al.: GIM^{3}E: condition-specific models of cellular metabolism developed from metabolomics and expression data. Bioinformatics **29**(22), 2900–8 (2013). https://doi.org/10.1093/bioinformatics/btt493. (Oxford, England)
18. Schultz, A., Qutub, A.A.: Reconstruction of tissue-specific metabolic networks using CORDA. PLoS Comput. Biol. (2016). https://doi.org/10.1371/journal.pcbi.1004808
19. Uhlen, M., Oksvold, P., Fagerberg, L., et al.: Towards a knowledge-based human protein Atlas. Nat. Biotechnol. (2010). https://doi.org/10.1038/nbt1210-1248
20. Vlassis, N., Pacheco, M.P., Sauter, T.: Fast reconstruction of compact context-specific metabolic network models. PLoS Comput. Biol. **10**(1), e1003424 (2014). https://doi.org/10.1371/journal.pcbi.1003424
21. Wang, Y., Eddy, J.A., Price, N.D.: Reconstruction of genome-scale metabolic models for 126 human tissues using mCADRE. BMC Syst. Biol. (2012). https://doi.org/10.1186/1752-0509-6-153
22. Yates, B., Bruford, E., Gray, K., et al.: Genenames.org: the HGNC and VGNC resources in 2019. Nucleic Acids Res. **47**(D1), D786–D792 (2018). https://doi.org/10.1093/nar/gky930
23. Zur, H., Ruppin, E., Shlomi, T.: iMAT: an integrative metabolic analysis tool. Bioinformatics (2010). https://doi.org/10.1093/bioinformatics/btq602

Deterministic Classifiers Accuracy Optimization for Cancer Microarray Data

Vânia Rodrigues[1] and Sérgio Deusdado[2(✉)]

[1] USAL – Universidad de Salamanca, 37008 Salamanca, Spain
[2] CIMO – Centro de Investigação de Montanha, Instituto Politécnico de Bragança, 5301-855 Bragança, Portugal
sergiod@ipb.pt

Abstract. The objective of this study was to improve classification accuracy in cancer microarray gene expression data using a collection of machine learning algorithms available in WEKA. State of the art deterministic classification methods, such as: Kernel Logistic Regression, Support Vector Machine, Stochastic Gradient Descent and Logistic Model Trees were applied on publicly available cancer microarray datasets aiming to discover regularities that provide insights to help characterization and diagnosis correctness on each cancer typology. The implemented models, relying on 10-fold cross-validation, parameterized to enhance accuracy, reached accuracy above 90%. Moreover, although the variety of methodologies, no significant statistic differences were registered between them, at significance level 0.05, confirming that all the selected methods are effective for this type of analysis.

Keywords: Classification · Cancer · Microarray · Datamining · Machine learning

1 Introduction

Accurate prediction and prognostic risk factor identification are essential to offer appropriate care for patients with cancer. Therefore, it is necessary to find biomarkers for the identification of different cancer typologies. Currently, with the evolution of microarray technology, it is possible for researchers to classify the types of cancer based on the patterns of gene activity in the tumor cells. For this purpose, statistical methods and machine learning techniques can be employed, such as classification methods to allow the assignment of class labels to samples with unknown biological condition, feature selection to identify informative genes and, additionally, clustering methods to discover classes of related biological samples. Detailed reviews on the technology and statistical methods often used in microarray analysis are presented in [1–3]. The objective of this work was to employ machine learning algorithms to analyze and classify gene expression data from cancer tissue samples provided by microarrays. The developed work included the use of three publicly available gene microarray datasets, described in the methodology section, on which the methods were tested and the performance assessed in order to compare the results with the best achievements published in the literature.

F. Fdez-Riverola et al. (Eds.): PACBB 2019, AISC 1005, pp. 154–163, 2020.
https://doi.org/10.1007/978-3-030-23873-5_19

This paper has been structured as follows. After a brief introduction, Sect. 2 describes the context and the state of art. Section 3 explains the methodology followed in this study, the procedures, the gene microarray datasets, the classification methods implemented as well as the optimal parameters adopted, concluding with the performance assessment of the classification methods. Experimental work using WEKA datamining workbench, the obtained results are discussed in Sect. 4 and the conclusions are presented in Sect. 5.

2 Background

2.1 Microarray Technology

In the last two decades microarrays were widely used to study gene expression. Main microarray technology includes Affymetrix [4] and Illumina [5] platforms. Other important microarray manufacturers are Exiqon [6], Agilent [7] or Taqman [8]. Gene microarray technology rest on the ability to deposit many (tens of thousands) different DNA sequences on a small surface, often referred to as a "chip". The different DNA fragments are arranged in rows and columns, in order to identify the location and distinguish the level of expression of each fragment on the array. Microarrays allow the measurement at expression level of a large simultaneous number of genes. Initially, the gene expression values are obtained by means of microscopic DNA spots attached to a solid surface which have followed a hybridization process [9], then it is possible to read the expression values with a laser, and subsequently store the quantification levels in a file.

Microarray technology has been extensively used by the scientific community. Accordingly, there has been a lot of data generation related to gene expression. This data is scattered and not easily available for public use. The National Center of Biotechnology Information (NCBI) organized, integrated and made available microarray data through a web service, the Gene Expression Omnibus or GEO. GEO is a data repository facility which includes data on gene expression from various sources.

Microarray technology possesses extensive applications in the medical field, mainly regarding diagnostics and prognostics. In this context, microarrays are widely used to know the state of a disease, type of tumor and other important factors for the patient treatment [10].

Considering disease diagnosis, it allows researchers to study and gather knowledge about many diseases such as mental illness, heart diseases, infectious disease and particularly the study of cancer [11]. They are also used in pharmacology response, which consists of the study of correlations between therapeutic responses to drugs and the genetic profiles of the patients, and additionally in the toxicological research to establish a correlation between responses to toxicants and the changes in the genetic profiles of the cells exposed to toxicants [12].

2.2 Deterministic Classifiers Overview

Kernel Logistic Regression (KLR) model is a statistical classifier [13] that generates a fit model by minimizing the negative log-likelihood with a quadratic penalty using the Broyden-Fletcher-Goldfard-Shanno (BFGS) optimization [14].

Support Vector Machine (SVM) algorithm is a discriminative classifier that tries to find an optimal hyperplane with maximal margin [15, 16]. SVM was developed for binary classification problems, although extensions to the technique have been made to support multi-class classification and regression problems [9]. This classifier is a state of the art classification system. In Cao et al. [17] SVM was applied in two-class datasets (Leukemia and colon Tumor) and also in multi-class datasets, proposing a novel fast feature selection method based on multiple SVDD (Support Vector Data Description). [18] focused on supervised gene expression analysis of cancers microarrays: prostate cancer, lymphoma and breast cancer. SVM algorithm is implemented in practice using selectable kernel functions. The kernel defines the similarity or a distance measure between new data and support vectors. The dot product is the similarity measure used for linear SVM or a linear kernel because the distance is a linear combination of the inputs. Other kernels can be used to transform the input space into higher dimensions such as Polynomial Kernel and a Radial Kernel. WEKA includes a derivative of SVM, the SMO implementation using sequential minimal optimization, described in [19].

The Stochastic Gradient Descent (SGD) algorithm implements a plain stochastic gradient descent learning routine which supports different loss functions and penalties for classification. Available loss functions include the Hinge loss (linear support vector classifier), Log loss (logistic regression), Squared loss (least squares linear regression), Epsilon-insensitive loss (support vector regression) and Huber loss (robust regression).

Decision tree classifiers recursively partition the instance space using hyperplanes that are orthogonal to axes. The model is built from a root node which represents an attribute and the instance space split is based on function of attribute values (split values are chosen differently for different algorithms), most frequently using its values. Then, each new sub-space of the data is split into new sub-spaces iteratively until an end criterion is met and the terminal nodes (leaf nodes) are each assigned a class label that represents the classification outcome (the class of all or majority of the instances contained in the sub-space). Setting the right end criterion is very important because trees that are too large can be overfitted and small trees can be underfitted, suffering a loss in accuracy in both cases. Most of the algorithms have a mechanism built in that deals with overfitting; it is called pruning.

Each new instance is classified by navigating them from the root of the tree down to a leaf, according to the outcome of the tests along the path [20, 21].

Although there are several methodologies to implement decision tree classifiers, for instance: SimpleCart, BFTree, FT, J48, LADTree, LMT and REPTree, the literature refers Logistic Model Trees (LMT) as the most efficient to classify microarray datasets [22].

A Logistic Model Trees (LMT) is a classification algorithm that integrates decision tree induction with logistic regression, building the logistic regression (LR) models at the leaves by incrementally refining those constructed at higher levels in the tree [23].

In the logistic variant, the LogitBoost algorithm [24] is used to produce an LR model at every node in the tree; the node is then split using the C4.5 criterion. Boosting works by sequentially applying a classification algorithm to reweighted versions of the training data and then taking a weighted majority vote of the sequence of classifiers thus produced. For many classification algorithms, this simple strategy results in a dramatic improvement in performance.

3 Methods

3.1 Experimental Procedures

The experimental work was based on the WEKA, version 3.8.3, a datamining workbench publicly accessible at: www.cs.waikato.ac.nz/ml/weka/. After data preparation and method selection (considering accuracy above 90%), using the explorer module, the module experimenter was used to automate experiments to achieve multiple classifiers comparison, testing with Paired T-Tester (Corrected). Prior to the experimental analysis, the microarray datasets were pre-processed and normalized on the interval [0, 1]. Successively, an external ten-fold cross-validation was performed, which randomly divides each dataset into ten equal parts. In each validation, one of them is taken as the testing set, and the others nine parts are used as the training set. The training and test data do not overlap each other to assure an unbiased comparison. Three functions based classifiers (KLR, SMO and SGD algorithms) and one decision tree classifier (LMT algorithm) were used as base learners. To compare classification performance, we created an experiment that ran 10 times several schemes (all classifications methods used) against each dataset with 10-fold cross-validation. Subsequently, we used literature mining analysis results to compare the performance of the methods applied in these microarrays datasets. These set of experiments were conducted on a computer with an Intel Core i7-5500U CPU 2.40 GHz processor, with 8.00 GB RAM.

3.2 Datasets

Three publicly available microarray datasets from different cancer typologies were used to test the classification methods, namely Leukemia, Lymphoma and Prostate datasets. The Leukemia datasets were obtained online from http://portals.broadinstitute.org/cgi-bin/cancer/publications/pub_paper.cgi?mode=view&paper_id=63, and were published as part of the experimental work in [25]. The Lymphoma and Prostate datasets were obtained online from http://ico2s.org/datasets/microarray.html, and were published as part of the experimental work in [26, 27].

All of them are two-class datasets. In Leukemia (a) dataset are present two types of leukemia: Acute Lymphoblastic (ALL) and Acute Myeloid Leukemia (AML). The leukemia dataset was analyzed in two different versions, the original composed by 52 samples and 12582 genes and a reduced version, composed by 28 samples keeping the same features. The goal for this subdivision was to test if the number of samples influences the prediction results.

Lymphoma dataset consists of 58 Diffuse large B-cell lymphoma samples vs. 19 follicular lymphoma samples.

Prostate cancer datasets consist of 52 tumor samples vs. 50 controls.

The composition details of the used datasets are shown below in Table 1.

Table 1. Used datasets characterization.

Dataset	2 classes	Genes	References
Leukemia(a)	24 – 28 (ALL–AML)	12582	[25]
Leukemia(b)	14 – 14 (ALL–AML)	12582	[25]
Diffuse large B-cell lymphoma	58 – 19	2647	[26]
Prostate cancer	52 – 50	2135	[27]

3.3 Classification Methods Parameterization

KLR Method

The parameters optimized were support vector with different types of kernel function. The penalty parameter λ with smaller values conjugated with different types of kernel functions was tested. The linear kernel function with $\lambda = 0.001$ presented the smaller mean absolute error.

SVM Method

We used the SMO classifier, a specific efficient optimization algorithm used to enhance the SVM performance. The model contains the complexity parameter C that influences the number of support vectors, we set C to 0.5. If C is lower, the more sensitive the algorithm becomes to training data, leading to higher variance and lower bias. With a higher C, the algorithm becomes less sensitive to the training data, in this case we obtain lower variance and higher bias. We tested polynomial functions of different degrees with different filters types without good results and, consequently a polynomial kernel without filter was selected, having set the exponent to 0.5.

SGD Method

SGD is an optimization method for unconstrained optimization problems. It approximates the true gradient by considering a single training example at a time. The algorithm works iteratively over the training examples and for each example updates the model parameters. The learning rate parameter was optimized setting a small value (0.0001) affecting the learning binary class SVM.

LMT Method

LMT consists of a tree structure that is made up of a set of inner or non-terminal nodes N and a set of leaves or terminal nodes T. Considering S the whole instance space, spanned by all attributes that are presented in the dataset. Then the tree structure gives a disjoint subdivision of S into regions S_t, and every region is represented by a leaf in the tree:

$$S = \bigcup_{t \in T} S_t \quad S_t \cap S_{t'} = \emptyset \text{ for } t \neq t'$$

The model represented by whole LMT is given by $F_j(x) = \alpha_0^j + \sum_{k=1}^{m} \alpha_{v_k}^j . v_k$.

If $\alpha_{v_k}^j = 0$ for $v_k \notin V_t$. The model of LMT is then given by

$$f(x) = \sum_{t \in T} f_t(x) \cdot I(x \in S_t)$$

Where $I(x \in S_t)$ is 1 if $x \in S_t$ and 0 otherwise. Considering the WEKA implementation of LMT, we used the fast regression heuristic that avoids cross-validating the number of LogitBoost iterations at every node [23]. LMT employs the minimal cost-complexity pruning mechanism to produce a compact tree structure.

3.4 Performance Evaluation

In this study, we trained the classifiers to predict outcomes of cancer microarray datasets contained positive samples and control samples. The evaluation measures to evaluate the classifiers [28, 29], include classification accuracy (*ACC*), *i.e.*, the ratio of the true positives and true negatives obtained by the classifier over the total number of instances in the test dataset, defined as:

$$ACC = \frac{TN + TP}{TP + FP + FN + TN}$$

Kappa (*k*) coefficient is a statistical measure for qualitative (categorical) items as given by:

$$k = \frac{Observed\ Accuracy - Expected\ Accuracy}{1 - Expected\ Accuracy}$$

Kappa coefficient is interpreted using the guidelines outlined by Landis and Koch (1977), where strength of the *k* is interpreted in the flowing manner: 0.01–0.20 slight; 0.21–0.40 fair; 0.41–0.60 moderate; 0.61–0.80 substantial; 0.81–1.00 almost perfect [30].

Mean Absolute Error (MAE) measures the average magnitude of the errors in a set of prediction, without considering their direction [31]. It is given by:

$$MAE = \frac{\sum_{i=1}^{n} |predicted_i - \text{actual}_i|}{total\ predictions}$$

Precision (*PRE*), it is also called the Positive Predictive Values (PPV), is the proportion of the true positives against the true positives and false positives, as given by equation:

$$PRE = \frac{TP}{TP + FP}$$

Recall (*REC*) also called sensitivity and hit rate, is the proportion of the true positives against true positives and false negatives, as given by the equation:

$$REC = \frac{TP}{TP + FN}$$

F-measure, it is also called F score, is the harmonic mean of precision and recall which is given by the equation:

$$f_{measure} = \frac{2 * PRE * REC}{PRE + REC}$$

ROC stands for Receiver Operating Characteristic. It's created by plotting the True Positives rates vs False Positives rates. It is also exploited to evaluate the performance of classifiers as Area Under ROC.

4 Results and Discussion

For each dataset in this study, the results of the classifiers estimation performance are presented in Table 2. These results are expressed on average, considering the 10 times that each test was repeated.

Table 2. Results achieved by algorithms with 10-fold cross-validation.

Dataset	Classifier	ACC (%) (st. dev.)	k (st. dev.)	MAE (st. dev.)	Recall (st. dev.)	F-Measure (st. dev.)	Area Under ROC (st. dev.)
Leukemia (a)	KLR	100	1	0.01 (0.01)	1	1	1
	SVM	100	1	0.00	1	1	1
	LMT	97.33 (6.67)	0.94 (0.14)	0.13 (0.08)*	0.94 (0.14)	0.96 (0.09)	1
	SGD	100	1	0.00	1	1	1
Leukemia (b)	KLR	98.17 (8.17)	0.95 (0.20)	0.02 (0.06)	1	0.99 (0.06)	1
	SVM	96.67 (10.86)	0.93 (0.24)	0.03 (0.11)	1	0.97 (0.09)	0.96 (0.12)
	LMT	100	1	0.14 (0.04)*	1	1	1
	SGD	96.33 (11.26)	0.92 (0.26)	0.04 (0.11)	1	0.97 (0.09)	0.96 (0.13)
Diffuse large B-cell lymphoma	KLR	95.50 (6.90)	0.87 (0.22)	0.05 (0.06)	0.97 (0.07)	0.97 (0.04)	0.98 (0.05)
	SVM	98.70 (3.94)	0.97 (0.09)	0.01 (0.04)	0.98 (0.05)	0.99 (0.03)	0.99 (0.03)
	LMT	92.25 (10.35)	0.77 (0.31)	0.09 (0.09)	0.96 (0.09)	0.95 (0.07)	0.94 (0.15)
	SGD	98.20 (4.84)	0.96 (0.11)	0.02 (0.05)	0.98 (0.06)	0.99 (0.04)	0.99 (0.04)
Prostate cancer	KLR	89.18 (8.61)	0.78 (0.17)	0.11 (0.08)	0.89 (0.13)	0.89 (0.09)	0.96 (0.06)
	SVM	92.33 (8.27)	0.85 (0.17)	0.08 (0.08)	0.96 (0.09)	0.93 (0.08)	0.92 (0.08)
	LMT	90.76 (8.83)	0.81 (0.18)	0.15 (0.07)	0.92 (0.12)	0.91 (0.09)	0.95 (0.07)
	SGD	90.18 (8.21)	0.80 (0.16)	0.10 (0.08)	0.93 (0.11)	0.90 (0.08)	0.90 (0.08)*

*Statistically different at significance level 0.05

The experiment was configured using KLR as the referential for all datasets, the results registered in Table 2 correspond to the comparison between the different classifiers considering the used evaluation measures.

Leukemia

On leukemia (a) dataset, the prediction results of KLR, SVM, and SGD are 100% ACC followed by LMT with ACC of ≈97%. Kappa coefficient results of KLR, SVM and SGD indicates a perfect agreement (1) between the classification and the true classes, having a LMT result almost perfect (≈0.94). F-measure and Area under ROC presents results nearly 1 on all methods, which indicates the good performance of the classification models used. These results are similar because there are not differences statistically significant between them. On the contrary, MAE is statistically better in KLR than LMT but not statistically significant differences on SVM and SGD, the same is verified on leukemia (b) dataset. On the leukemia (b) dataset, LMT achieves 100% ACC. On the contrary, SVM and SGD achieved ACC ≈ 96.67% and ACC ≈ 96.33%, respectively, but they do not present differences statistically significant as well. In comparison, the cross-validation results reported in literature for this datasets [17], presented results of SVM methods using kernel functions achieving results of average recall equal to 93.93%. In the cited work, the authors optimized the method to achieve the best results equal to 96.43% of average recall, however their study used a smaller number of features. In [32], leukemia datasets with smaller number of features presented the maximum ACC results equal to 97.43% using the RBF Network classifier.

Diffuse Large B-cell Lymphoma

For the Diffuse Large B-cell lymphoma dataset, the ACC result of SVM is 98.70% and SGD is 98.20%, followed by KLR with 95.50% and then LMT with 92.25%. K results of LMT revealed a substantial agreement (0.77) between the classifications and the true classes, whereas the other classifiers presented almost a perfect agreement. On the lymphoma dataset there are no statistical differences on MAE among the four classifiers. F-measure and Area under ROC indicates an excellent prediction of the classification methods (≥ 0.9). In literature, the results reported for this dataset [18] achieved 95% ACC using a higher number of features. The same datasets were analyzed in [32] and presented the best outcome prediction having ACC equal to 92.45%. In [33] was achieved 95.7% ACC, also with high number of features.

Prostate Cancer

For the prostate cancer dataset, the best ACC result was 92.33% and was obtained with SVM. LMT, SGD and KLR achieved very close results, respectively 90.76%, 90.18% and 89.18%. KLR and SGD Kappa coefficient results indicate substantial agreement ≈0.78, 0.80, respectively, between these classifiers and the true classes, while in SVM and LMT presented almost perfect agreement, with k equal to 0.85 and 0.81, respectively. However, all k do not present differences statistically significant. Analyzing the results of Area under ROC, there is a significant statistical difference between KLR (0.96) and SGD (0.90). On the contrary, LMT (0.95) and SVM (0.92) are not statistically different. F-measure results were very close to 1, which means good performance of all classifiers implemented. Comparatively with our work, in [18] was used a higher number of features in the cross-validation results for this dataset, achieving 94% ACC. In the papers published by [34] and [33] were obtained 94.6% and 93.4% ACC, respectively. In [32] the best outcome prediction measured by ACC was equal to 95.20% using the SVM classifier.

5 Conclusions

All the classifiers involved in this study (KLR, SVM, LMT, SGD) presented good performance in gene expression analysis on cancer microarrays data, proving to be effective and reliable in this type of data. The classifiers performance, except for the measures MAE and Area under ROC, in some schemes, are not statistically different. The developed experimental work achieved better or close-to-best performance by comparison with other methods applied on the same datasets in the literature.

References

1. Allison, D.B., Cui, X., Page, G.P., Sabripour, M.: Microarray data analysis: from disarray to consolidation and consensus. Nat. Rev. Genet. **7**, 55–65 (2006)
2. Hoheisel, J.D.: Microarray technology: beyond transcript profiling and genotype analysis. Nat. Rev. Microbiol. **7**, 200–210 (2006)
3. Quackenbush, J.: Computational analysis of microarray data: computational genetics. Nat. Rev. Genet. **2**, 418–427 (2001)
4. Talloen, W., Göhlmann, H.: Gene Expression Studies Using Affymetrix Microarrays. Chapman and Hall/CRC (2009)
5. Illumina: Illumina Genes Expression arrays (2009)
6. Exiqon: Exiqon Genes Expression arrays (2009)
7. Zahurak, M., Parmigiani, G., Yu, W., Scharpf, R.B., Berman, D., Schaeffer, E., Shabbeer, S., Cope, L.: Pre-processing agilent microarray data. BMC Bioinform. **8**, 142 (2007)
8. Taqman: Taqman Genes Expression arrays (2009)
9. Castillo, D., Gálvez, J.M., Herrera, L.J., Román, B.S., Rojas, F., Rojas, I.: Integration of RNA-Seq data with heterogeneous microarray data for breast cancer profiling. BMC Bioinform. **18**, 506 (2017)
10. Kaliyappan, K., Palanisamy, M., Govindarajan, R., Duraiyan, J.: Microarray and its applications. J. Pharm. Bioallied Sci. **4**, 310 (2012)
11. Raghavachari, N.: Microarray technology: basic methodology and application in clinical research for biomarker discovery in vascular diseases. In: Freeman, L.A. (ed.) Lipoproteins and Cardiovascular Disease, pp. 47–84. Humana Press, Totowa (2013)
12. Scherf, U., Ross, D.T., Waltham, M., Smith, L.H., Lee, J.K., Tanabe, L., Kohn, K.W., Reinhold, W.C., Myers, T.G., Andrews, D.T., Scudiero, D.A., Eisen, M.B., Sausville, E.A., Pommier, Y., Botstein, D., Brown, P.O., Weinstein, J.N.: A gene expression database for the molecular pharmacology of cancer. Nat. Genet. **24**, 236–244 (2000)
13. Wahba, G., Gu, C., Wang, Y., Chappell, R.: Soft classification, A.K.A. risk estimation, via penalized log likelihood and smoothing spline analysis of variance. In: Computational Learning Theory and Natural Learning Systems, pp. 133–162. MIT Press (1995)
14. Smith, B., Wang, S., Wong, A., Zhou, X.: A penalized likelihood approach to parameter estimation with integral reliability constraints. Entropy **17**, 4040–4063 (2015)

15. Boser, B.E., Guyon, I.M., Vapnik, V.N.: A training algorithm for optimal margin classifiers. In: Proceedings of the Fifth Annual Workshop on Computational Learning Theory - COLT 1992, pp. 144–152. ACM Press, Pittsburgh (1992)
16. Vapnik, V.N.: Statistical Learning Theory. Wiley, New York (1998)
17. Cao, J., Zhang, L., Wang, B., Li, F., Yang, J.: A fast gene selection method for multi-cancer classification using multiple support vector data description. J. Biomed. Inform. **53**, 381–389 (2015)
18. Glaab, E., Bacardit, J., Garibaldi, J.M., Krasnogor, N.: Using rule-based machine learning for candidate disease gene prioritization and sample classification of cancer gene expression data. PLoS ONE **7**, e39932 (2012)
19. Schölkopf, B., Burges, C.J.C., Smola, A.J. (eds.): Advances in Kernel Methods: Support Vector Learning. MIT Press, Cambridge (1999)
20. Polaka, I., Tom, I., Borisov, A.: Decision tree classifiers in bioinformatics. Sci. J. Riga Tech. Univ. Comput. Sci. **42**, 118–123 (2010)
21. Rokach, L., Maimon, O.: Data Mining with Decision Trees: Theory and Applications. World Scientific, Hackensack (2015)
22. Li, Y., Wang, N., Perkins, E.J., Zhang, C., Gong, P.: Identification and optimization of classifier genes from multi-class earthworm microarray dataset. PLoS ONE **5**, e13715 (2010)
23. Landwehr, N., Hall, M., Frank, E.: Logistic model trees. Mach. Learn. **59**, 161–205 (2005)
24. Friedman, J., Hastie, T., Tibshirani, R.: Additive logistic regression: a statistical view of boosting (with discussion and a rejoinder by the authors). Ann. Stat. **28**, 337–407 (2000)
25. Armstrong, S.A., Staunton, J.E., Silverman, L.B., Pieters, R., den Boer, M.L., Minden, M. D., Sallan, S.E., Lander, E.S., Golub, T.R., Korsmeyer, S.J.: MLL translocations specify a distinct gene expression profile that distinguishes a unique leukemia. Nat. Genet. **30**, 41–47 (2001)
26. Shipp, M.A., Ross, K.N., Tamayo, P., Weng, A.P., Kutok, J.L., Aguiar, R.C.T., Gaasenbeek, M., Angelo, M., Reich, M., Pinkus, G.S., Ray, T.S., Koval, M.A., Last, K.W., Norton, A., Lister, T.A., Mesirov, J., Neuberg, D.S., Lander, E.S., Aster, J.C., Golub, T.R.: Diffuse large B-cell lymphoma outcome prediction by gene-expression profiling and supervised machine learning. Nat. Med. **8**, 68–74 (2002)
27. Singh, D., Febbo, P.G., Ross, K., Jackson, D.G., Manola, J., Ladd, C., Tamayo, P., Renshaw, A.A., D'Amico, A.V., Richie, J.P., Lander, E.S., Loda, M., Kantoff, P.W., Golub, T.R., Sellers, W.R.: Gene expression correlates of clinical prostate cancer behavior. Cancer Cell **1**, 203–209 (2002)
28. Saito, T., Rehmsmeier, M.: The precision-recall plot is more informative than the ROC plot when evaluating binary classifiers on imbalanced datasets. PLoS ONE **10**, e0118432 (2015)
29. Tharwat, A.: Classification assessment methods. Appl. Comput. Inform. (2018). https://doi.org/10.1016/j.aci.2018.08.003
30. Landis, J.R., Koch, G.G.: The measurement of observer agreement for categorical data. Int. Biom. Soc. **33**, 159–174 (1977)
31. Sammut, C., Webb, G.I. (eds.): Encyclopedia of Machine Learning. Springer, Boston (2010)
32. Dagliyan, O., Uney-Yuksektepe, F., Kavakli, I.H., Turkay, M.: Optimization based tumor classification from microarray gene expression data. PLoS ONE **6**, e14579 (2011)
33. Wessels, L.F.A., Reinders, M.J.T., Hart, A.A.M., Veenman, C.J., Dai, H., He, Y.D., van't Veer, L.J.: A protocol for building and evaluating predictors of disease state based on microarray data. Bioinformatics **21**, 3755–3762 (2005)
34. Shen, L., Tan, E.C.: Dimension reduction-based penalized logistic regression for cancer classification using microarray data. IEEE/ACM Trans. Comput. Biol. Bioinform. **2**, 166–175 (2005)

Artificial Intelligence in Biological Activity Prediction

João Correia(✉), Tiago Resende, Delora Baptista, and Miguel Rocha

CEB - Centre of Biological Engineering, University of Minho, Campus of Gualtar, Braga, Portugal
jfscorreia95@gmail.com, tiagofcresende@gmail.com, dlr.baptista@gmail.com, mrocha@di.uminho.pt

Abstract. Artificial intelligence has become an indispensable resource in chemoinformatics. Numerous machine learning algorithms for activity prediction recently emerged, becoming an indispensable approach to mine chemical information from large compound datasets. These approaches enable the automation of compound discovery to find biologically active molecules with important properties. Here, we present a review of some of the main machine learning studies in biological activity prediction of compounds, in particular for sweetness prediction. We discuss some of the most used compound featurization techniques and the major databases of chemical compounds relevant to these tasks.

Keywords: Machine learning · Deep learning · Biological activity prediction · Sweetness prediction · Compound featurization

1 Introduction

For centuries, humans have been manually searching and documenting different compounds, assessing their interaction with biological systems to find suitable products that solve problems and enhance quality of life. Despite the broad amount of data collected on compounds capable of curing illnesses, fighting infections or satisfying our food sensory system, the search for compounds with improved biological capabilities is still in high demand. With the modernization of the pharmaceutical and food industries, there is a growing need for more sustainable compounds, with improved biological activities. By taking advantage of the enormous quantity of categorized data on compound biological activity existent, and still being generated, new approaches using artificial intelligence (AI) are continuously being developed. With datasets getting larger and more detailed and algorithms increasing their scope and accuracy, new tools for predicting biological activity arise, accelerating the generation of new products.

F. Fdez-Riverola et al. (Eds.): PACBB 2019, AISC 1005, pp. 164–172, 2020.
https://doi.org/10.1007/978-3-030-23873-5_20

2 Artificial Intelligence in Biological Activity Prediction

Machine learning (ML) is a field of AI where systems learn from data, identify patterns and make decisions without being explicitly programmed [1]. Although ML algorithms were created in the 1950s [2], ML only started to thrive in the 1990s and is becoming the most popular sub-field of AI. ML techniques are classified as supervised or unsupervised. In the former, given input-output pairs, a function to map the input to the output is learned so the model can predict future cases. In the latter, patterns are learned directly from unlabeled data. For biological activity prediction, it is common to use supervised methods.

In linear regression (LR) and logistic regression (LgR), linear relationships between independent and dependent variables are learned. LgR is used for linear classification when the dependent variable is categorical. Naive Bayess (NBs) is a probabilistic classification algorithm based on the Bayes theorem and the assumption of feature independence. Random forests (RFs) are ensembles of decision trees (DTs), tree-like models of decision rules where each node represents a feature, each branch represents a decision and each leaf an outcome. RFs apply bagging to generate distinct training sets and create different models, and predictions are obtained by majority voting. The objective of support-vector machines (SVMs) is to map the data into a high-dimensional space by identifying a lower dimensional hyperplane that separates the data using nonlinear kernels. K-nearest neighborss (KNNs) is an instance-based algorithm where data is classified by its similarity with its k-nearest neighbors. Partial least squares (PLS) regression is mostly used to predict a set of dependent variables from a large set of independent variables. PLS decomposes the original set of variables into a set of components that explain the most covariance between the independent and dependent variables, and uses these components to predict the outputs. Neural networks (NNs) are biologically-inspired algorithms designed to automatically recognize patterns from input labeled data in order to be able to predict the output of unlabeled data according to similarities with the example inputs.

Due to high accuracy and cost-effectiveness, ML is extensively used in many fields including chemoinformatics. Recent algorithmic advances, as well as the development of databases for the storage of molecule structures and their properties, accelerated the pace at which the field has evolved. Researchers have used combinations and different approaches of traditional ML, as well as complex deep learning (DL) architectures. A common approach is the use of these models for the optimization of quantitative structure-activity relationship (QSAR) models to improve the biological activity prediction of multiple compounds.

Biological activity prediction of compounds is one of the main research areas in chemoinformatics [3]. The application of AI for this type of task is of critical importance for the identification of compounds with desired properties. The objective is to select a subset of compounds from all the compounds under consideration that have a higher probability of being bioactive when compared to a random sample. One important aspect for the success of ML in property prediction is the access to large datasets. Multiple large datasets from public-domain repositories are available and suited for activity prediction; such is the

Table 1. A selection of recent studies that use AI for biological activity prediction

ML methods	Study description
SVMs, KNN, RFs, NB, DNNs	Comparison of DL methods on a large-scale drug discovery dataset and other ML and target prediction methods [10]
RF, KNN, NB, DNN	Predicting kinase activities for around 200 different kinases using multiple ML methods [11]
NB, SVMs, LgR, RFs, DNN	Different ML methods were compared using a standardized dataset from ChEMBL [13] and standardized metrics [14]
NB, LgR, DTs, RFs, SVMs, DNNs	Comparison between DNNs and other ML algorithms for diverse end-points (bioactivity, solubility and ADME properties) [15]
Multitask DNNs	Use of multitask DNNs as an improvement over single task learning [16]
DNNs, SVMs, RFs, NB, KNN	Shows that, when optimized, DNNs are capable of outperforming shallow methods across diverse activity classes [17]
DNNs, SVMs and RFs	Results from the Tox21 competition. DNNs show good predictivity on 12 different toxic endpoints [18]
Multitask DNNs	Multitask learning provided benefits over single task models. Smaller datasets tend to benefit more than larger datasets [19]
Multitask DNNs	Performance analysis of multitask DNNs (DeepChem implementation) and related DL models on pharmaceutical datasets [20]
DNNs, RFs, DTs	Comparison between multitask DNNs and alternative ML methods. Multitask DNNs outperformed alternative methods [21]
DNNs, RFs	Performance comparison between DNNs and RFs for QSARs using different datasets and descriptors [12]
DNNs	DL models to predict drug-induced liver injury [22]
Multitask DNNs	Multitask vs Single Task learning. Multitask DNNs showed better performance [23]
DNNs, SVMs, LgR, KNN, etc	Comparison of DL performance against multiple ML methods using data from ChEMBL [24]
NB and RFs	Comparison between NB and RFs to make accurate ADME-related activities predictions on 18 large QSAR datasets [25]
Shallow NNs	Prediction of biological activities of structurally diverse ligands using 3 types of fingerprints (ECFP6, FP2 & MACCS) [26]
Bayesian QSAR, PLS	Bayesian QSAR combines activities across the kinase family to predict affinity, selectivity, and cellular activity [27]

case of DrugBank [4], PubChem BioAssay [5], ChEBI [6], MoleculeNet datasets [7], ChemSpider [8] and T3DB: the toxic exposome database [9].

ML techniques, including SVMs, RFs and deep neural networks (DNNs) have been used to discover compounds with desired biological activities. Table 1 summarizes some of the most relevant studies in biological activity prediction using AI. In general, DNNs exhibit better results than classic ML methods [10,11], where RF-based models are the most used and the ones showing better results, even outperforming DNNs in some situations [12].

2.1 Compound Featurization

Information for biological activity prediction comes primarily from the chemical structure of the compound. There has been a lot of research on how to transform molecules into a form suited for ML algorithms so that the model can learn and generalize the properties shared among the molecules. Some of the most useful molecular featurization methods include line notations, fingerprints, weave, graph convolutions and (NLP)-inspired embeddings.

Line Notations. Line notations express the 2D structure of compounds. These approaches represent structures and chemical properties such as atoms, bonds, aromaticity, chiral and isotopic information of compounds as compact strings of characters [28]. The most common line notations are SMILES [29] and InChI [30]. Most chemical databases like PubChem [5], ChEMBL [13] and DrugBank [4] provide line notations for the recorded compounds.

Fingerprints. Fingerprints are the most widely used molecular representations in chemoinformatics. They consist of binary arrays, where each dimension represents the presence or absence of a particular substructure or property. Fingerprints are used to encode multiple characteristics, including atomic attributes, atomic environments, bond properties and bond positions which enables it to be applied to tasks such as activity prediction. Extended-Connectivity Fingerprints (ECFPs), Functional-Class Fingerprints (FCFPs), and the 166-bit Molecular Access System (MACCS) are typical fingerprint-based featurization approaches.

Graph Convolutions. In this DL-based approach proposed by Duvenaud et al. [31], the chemical structure of the molecule is initially represented as a graph with atoms as nodes and bonds as edges, encoding the connectivity between atoms and each atom's local chemical environment. Then, different neighbor levels of the molecule graph representation are fed into a single layer convolutional NN to generate fixed-length vectors. The resulting vectors are transformed through a pooling-like operation using the softmax function, and then they are summed to form the final molecular-level vector representations.

Weave. The weave featurization method is very similar to graph convolutions. It also encodes both the connectivity between atoms in a molecule and each atom's local chemical environment, but connectivity uses more detailed pair features instead of information for the neighbor's list. It also encodes both the connectivity between atoms in a molecule and each atom's local chemical environment, but uses more detailed pair features instead of information for the neighbors list. Weave modules combine and transform the atom-level and pair-level features by applying specific convolution operators [32].

NLP-Inspired Embeddings. Deep learning-based NLP techniques can be directly applied to SMILES strings to generate continuous feature vectors instead of learning from molecular graphs. The Seq2seq fingerprint [33] translates molecules represented as SMILES strings into continuous embeddings using a model based on the sequence to sequence [34] machine translation model. Mol2vec [35] is a method inspired by the Word2vec [36] word embedding algorithm that learns continuous embeddings of compound substructures.

2.2 Sweetness Prediction

Sweetness prediction is a particular application of biological activity prediction, very important for many disciplines, especially food chemistry. As sugars and saccharides are widely used in the food industry, their overconsumption can severely affect human health, leading to serious diseases, such as obesity, diabetes and cardiovascular diseases. It is, thus, of extreme importance to identify low-calorie sweeteners present in natural or chemically synthesized compounds, avoiding, this way, associated health risks while preserving the sweetness perception.

The high cost associated with compound sweetness determination in the laboratory remains a barrier, justifying the necessity to build computational models capable of learning the relationship between sweetness and the structure of known sweeteners. Therefore, these models would facilitate the identification and design of new sweeteners with different degrees of sweetness. Moreover, existing sweeteners have been the subject of controversies regarding health and food safety [37]. In this aspect, computational methods for biological activity prediction can offer additional value by combining sweetness prediction with other tasks such as toxicity and bitterness prediction, removing in this way compounds with undesirable properties.

In recent years, multiple ML based models to predict compound sweetness were developed. In 2011, Yang et al. [38], developed three quantitative models (linear regression, neural networks (ANN), SVM) for the prediction of the sweetness of 103 compounds. Zhong et al. [39], in 2013, developed two quantitative models (linear regression and SVM) to predict the sweetness of 320 compounds. In 2016 and 2017, Rojas et al. [40,41] used KNN to discriminate sweet from non-sweet molecules. In the same year, Chéron et al. [42] used RF to predict either sweetness, bitterness and toxicity properties. In 2018, Goel et al. [43] developed QSAR models based on Genetic Function Approximation and ANNs analysis to predict the sweetness of molecules. A RF-based binary classifier to predict the bitterness and sweetness of chemical compounds was implemented by Banerjee et al. [44]. Ojha et al. [45] proposed 13 new sweet molecules using a quantitative structure-property relationship model and PLS regression analysis. In 2019, Zheng et al. [46] implemented multiple ML methods (KNN, SVM, Gradient Boosting Machine, RF, and DNN) for the prediction of sweeteners and their corresponding relative sweetness. Comparing the results obtained in the above mentioned studies is not completely feasible, because different datasets, number and type of descriptors and validation methods were used. However, a simple comparison shows that nonlinear methods such as RFs, SVMs, PLSs and

Table 2. Available databases containing data on sweeteners/non-sweeteners.

Database	Description
SweetenersDB [42] (http://chemosim.unice.fr/SweetenersDB/)	316 compounds belonging to 17 chemical families with known sweetness values
SuperSweet [47] (http://bioinformatics.charite.de/sweet/)	More than 15,000 natural and artificial sweeteners. Information on origin, sweetness class, predicted toxicity, molecular targets, etc.
FooDB (http://foodb.ca/)	The largest and most comprehensive database on food constituents
BitterDB [48] (http://bitterdb.agri.huji.ac.il/dbbitter.php)	Information on over 1,000 bitter-tasting natural & synthetic compounds
FlavorDB [49] (https://cosylab.iiitd.edu.in/flavordb/)	Contains 25,595 flavor molecules (618 sweet-tasting, 253 bitter-tasting)
Super natural II [50] (http://bioinf-applied.charite.de/supernatural_new/index.php)	Database comprising 325,508 natural compounds. Includes information about 2D structures, physicochemical properties and predicted toxicity

ANNs exhibit slightly better results. These methods, in general, can more easily capture the sweetness chemical space and therefore the structural diversity of known sweeteners, generating better results. Nonetheless, more accurate models are still in high demand. The use of DNNs models and taking into account the complex interactions between different sweeteners and respective receptors can further improve the results in the field.

With the generation of vast amounts of data from experimental and computational screening experiments, the need for structured databases to store and publish the generated data in a well-organized way is increasing. As a result, several compound databases that store thousands of molecules and respective chemical attributes, molecular descriptors, activity measurements and other information are available through the web. In particular, databases containing data on sweet/bitter molecules are starting to become more common. Table 2 describes the main databases containing sweet/non-sweet compounds.

3 Concluding Remarks

Here, we provide a review of literature related to AI algorithms used for biological activity prediction and in particular for sweetness prediction. Over the last decades, ML witnessed rapid development, and multiple methods have been successfully applied in chemoinformatics. Both shallow and DL methods have been widely used in this task and they have an important role in its future.

With the increase in the complexity and the size of the available datasets, DL models seem to frequently outperform traditional shallow ML algorithms. It is also common to benefit from multitask learning, as it has been shown that the prediction of related properties seems to be beneficial to the predictive performance of the models. The use of AI in chemoinformatics strongly benefits from open source implementations of different ML models and from the availability of extensive datasets allowing the implementation of fine-tuned complex NNs. With the progress of AI in chemoinformatics, an increase in the use of these approaches to automate compound discovery is expected.

With this review, we can conclude that improved methods are still in high demand. Combining state-of-the-art deep learning models with different data types and with approaches from different fields will be crucial for the discovery of added-value compounds. Following this research line, we are implementing in our group methods to improve the identification and generation of new sweeteners that can be produced using only biologically feasible reactions, replacing the chemical synthesis currently used.

Acknowledgments. This study was supported by the European Commission through project *SHIKIFACTORY100 - Modular cell factories for the production of 100 compounds from the shikimate pathway* (Reference 814408), and by the Portuguese FCT under the scope of the strategic funding of UID/BIO/04469/2019 unit and BioTecNorte operation (NORTE-01-0145-FEDER-000004) funded by the European Regional Development Fund under the scope of Norte2020.

References

1. Murphy, K.P.: Machine Learning: A Probabilistic Perspective. MIT Press, Cambridge [u.a.] (2013)
2. Samuel, A.L.: Some studies in machine learning using the game of checkers. IBM J. Res. Dev. **3**(3), 210–229 (1959)
3. Toccaceli, P., et al.: Conformal prediction of biological activity of chemical compounds. Ann. Math. Artif. Intell. **81**(1–2), 105–123 (2017)
4. Wishart, D.S., et al.: DrugBank 5.0: a major update to the DrugBank database for 2018. Nucleic Acids Res. **46**(D1), D1074–D1082 (2017)
5. Kim, S., et al.: PubChem 2019 update: improved access to chemical data. Nucleic Acids Res. **47**(D1), D1102–D1109 (2018)
6. Hastings, J., et al.: ChEBI in 2016: Improved services and an expanding collection of metabolites. Nucleic Acids Res. **44**(D1), D1214–D1219 (2015)
7. Wu, Z., et al.: MoleculeNet: a benchmark for molecular machine learning. Chem. Sci. **9**(2), 513–530 (2018)
8. Pence, H.E., Williams, A.: ChemSpider: an online chemical information resource. J. Chem. Educ. **87**(11), 1123–1124 (2010)
9. Wishart, D., et al.: T3DB: the toxic exposome database. Nucleic Acids Res. **43**(D1), D928–D934 (2014)
10. Mayr, A., et al.: Large-scale comparison of machine learning methods for drug target prediction on ChEMBL. Chem. Sci. **9**(24), 5441–5451 (2018)
11. Merget, B., et al.: Profiling prediction of kinase inhibitors: toward the virtual assay. J. Med. Chem. **60**(1), 474–485 (2016)
12. Ma, J., et al.: Deep neural nets as a method for quantitative structure-activity relationships. J. Chem. Inf. Model. **55**(2), 263–274 (2015)

13. Gaulton, A., et al.: The ChEMBL database in 2017. Nucleic Acids Res. **45**(D1), D945–D954 (2016)
14. Lenselink, E.B., et al.: Beyond the hype: deep neural networks outperform established methods using a ChEMBL bioactivity benchmark set. J. Cheminformatics **9**(1), 45 (2017)
15. Korotcov, A., et al.: Comparison of deep learning with multiple machine learning methods and metrics using diverse drug discovery data sets. Mol. Pharm. **14**(12), 4462–4475 (2017)
16. Xu, Y., et al.: Demystifying multitask deep neural networks for quantitative structure-activity relationships. J. Chem. Inf. Model. **57**(10), 2490–2504 (2017)
17. Koutsoukas, A., et al.: Deep-learning: investigating deep neural networks hyperparameters and comparison of performance to shallow methods for modeling bioactivity data. J. Cheminformatics **9**(1), 42 (2017)
18. Mayr, A., et al.: DeepTox: toxicity prediction using deep learning. Front. Environ. Sci. **3**, 80 (2016)
19. Kearnes, S., et al.: Modeling industrial ADMET data with multitask networks, June 2016
20. Ramsundar, B., et al.: Is multitask deep learning practical for pharma? J. Chem. Inf. Model. **57**(8), 2068–2076 (2017)
21. Dahl, G., Jaitly, N., Salakhutdinov, R.: Multi-task neural networks for QSAR predictions. CoRR arXiv:1406.1231v1 (2014)
22. Xu, Y., et al.: Deep learning for drug-induced liver injury. J. Chem. Inf. Model. **55**(10), 2085–2093 (2015)
23. Ramsundar, B., et al.: Massively multitask networks for drug discovery. CoRR arXiv:1502.02072 (2015)
24. Unterthiner, T., et al.: Deep learning as an opportunity in virtual screening, January 2014
25. Chen, B., et al.: Comparison of random forest and pipeline pilot naïve bayes in prospective QSAR predictions. J. Chem. Inf. Model. **52**(3), 792–803 (2012)
26. Myint, K.Z., et al.: Molecular fingerprint-based artificial neural networks QSAR for ligand biological activity predictions. Mol. Pharm. **9**(10), 2912–2923 (2012)
27. Martin, E., et al.: Profile-QSAR: a novel meta-QSAR method that combines activities across the kinase family to accurately predict affinity, selectivity, and cellular activity. J. Chem. Inf. Model. **51**(8), 1942–1956 (2011)
28. O'Boyle, N.M.: Towards a universal SMILES representation - a standard method to generate canonical SMILES based on the InChI. J. Cheminformatics **4**(1), 22 (2012)
29. Weininger, D.: SMILES, a chemical language and information system. 1. introduction to methodology and encoding rules. J. Chem. Inf. Model. **28**(1), 31–36 (1988)
30. Heller, S.R., et al.: InChI, the IUPAC international chemical identifier. J. Cheminformatics **7**(1), 23 (2015)
31. Duvenaud, D.K., et al.: Convolutional networks on graphs for learning molecular fingerprints. CoRR arXiv:1509.09292 (2015)
32. Kearnes, S., et al.: Molecular graph convolutions: moving beyond fingerprints. J. Comput.-Aided Mol. Des. **30**(8), 595–608 (2016)
33. Xu, Z., et al.: Seq2seq fingerprint. In: Proceedings of the 8th ACM International Conference on Bioinformatics, Computational Biology, and Health Informatics, ACM-BCB 2017, pp. 285–294. ACM Press, New York (2017)
34. Sutskever, I., et al.: Sequence to sequence learning with neural networks. In: Advances in Neural Information Processing Systems, pp. 3104–3112 (2014)

35. Jaeger, S., et al.: Mol2vec: unsupervised machine learning approach with chemical intuition. J. Chem. Inf. Model. **58**(1), 27–35 (2018)
36. Mikolov, T., et al.: Efficient estimation of word representations in vector space, January 2013
37. Whitehouse, C.R., et al.: The potential toxicity of artificial sweeteners. AAOHN J. **56**(6), 251–259 (2008)
38. Yang, X., et al.: In-silico prediction of sweetness of sugars and sweeteners. Food Chem. **128**(3), 653–658 (2011)
39. Zhong, M., et al.: Prediction of sweetness by multilinear regression analysis and support vector machine. J. Food Sci. **78**(9), S1445–S1450 (2013)
40. Rojas, C., et al.: A new QSPR study on relative sweetness. Int. J. Quant. Struct.-Prop. Relat. **1**(1), 78–93 (2016)
41. Rojas, C., et al.: A QSTR-based expert system to predict sweetness of molecules. Front. Chem. **5**, 53 (2017)
42. Chéron, J.B., et al.: Sweetness prediction of natural compounds. Food Chem. **221**, 1421–1425 (2017)
43. Goel, A., et al.: In-silico prediction of sweetness using structure-activity relationship models. Food Chem. **253**, 127–131 (2018)
44. Banerjee, P., Preissner, R.: BitterSweetForest: a random forest based binary classifier to predict bitterness and sweetness of chemical compounds. Front. Chem. **6**, 93 (2018)
45. Ojha, P.K., Roy, K.: Development of a robust and validated 2D-QSPR model for sweetness potency of diverse functional organic molecules. Food Chem. Toxicol. **112**, 551–562 (2018)
46. Zheng, S., et al.: e-sweet: a machine-learning based platform for the prediction of sweetener and its relative sweetness. Front. Chem. **7**, 35 (2019)
47. Ahmed, J., et al.: SuperSweet-a resource on natural and artificial sweetening agents. Nucleic Acids Res. **39**(Database), D377–D382 (2010)
48. Dagan-Wiener, A., et al.: Bitter or not? BitterPredict, a tool for predicting taste from chemical structure. Sci. Rep. **7**(1) (2017)
49. Garg, N., et al.: FlavorDB: a database of flavor molecules. Nucleic Acids Res. **46**(D1), D1210–D1216 (2017)
50. Banerjee, P., et al.: Super natural II–a database of natural products. Nucleic Acids Res. **43**(D1), D935–D939 (2014)

Towards the Reconstruction of Integrated Genome-Scale Models of Metabolism and Gene Expression

Fernando Cruz[1(✉)], Diogo Lima[1], José P. Faria[2], Miguel Rocha[1], and Oscar Dias[1(✉)]

[1] Centre of Biological Engineering, University of Minho, 4710-057 Braga, Portugal
{fernando.cruz, odias}@ceb.uminho.pt, diogolima95@hotmail.com, mrocha@di.uminho.pt

[2] Mathematics and Computer Science Division, Argonne National Laboratory, Argonne, IL, USA
jplfaria@anl.gov

Abstract. The reconstruction of integrated genome-scale models of metabolism and gene expression has been a challenge for a while now. In fact, various methods that allow integrating reconstructions of Transcriptional Regulatory Networks, gene expression data or both into Genome-Scale Metabolic Models have been proposed. Several of these methods are surveyed in this article, which allowed identifying their strengths and weaknesses concerning the reconstruction of integrated models for multiple prokaryotic organisms. Additionally, the main resources of regulatory information were also surveyed, as the existence of novel sources of regulatory information and gene expression data may contribute for the improvement of methodologies referred herein.

Keywords: Genome-Scale Metabolic Models · Genome-scale models of metabolism and gene expression · Regulation of gene expression · Databases of regulatory information

1 Control of Gene Expression in Prokaryotes

The optimal composition of the proteome in prokaryotes and eukaryotes changes considerably over time. In prokaryotic organisms, these changes often reflect the cell response to an ever-changing environment. Hence, the regulation of the gene expression is pivotal for controlling the optimal cellular composition of the proteome as a function of the consecutive environmental conditions.

The control of gene expression can occur at several potential stages of regulation [1]. Nevertheless, this review focuses primarily on the regulation of transcription initiation, as it likely is the main control stage of gene expression, in prokaryotic cells [2]. In addition, this study emphasizes the control of gene expression associated with the regulation of the cell metabolism.

Transcription is initiated when the holoenzyme RNA polymerase binds to a specific region of DNA known as the promoter [2]. Although there are consensus sequences for

F. Fdez-Riverola et al. (Eds.): PACBB 2019, AISC 1005, pp. 173–181, 2020.
https://doi.org/10.1007/978-3-030-23873-5_21

many promoters, these may vary considerably within the genome. As a result, the binding affinity of RNA polymerase is affected, and consequently the rate at which transcription is initiated. Due to this control over the initiation of transcription, these DNA sequences are often classified as strong or weak promoters.

Many of the principles assumed in the regulation of prokaryotic gene expression are based on the fact that genes are clustered into operons, which are regulated together [2]. In prokaryotes, genes are placed linearly and sequentially. A single functional mRNA molecule contains the information for the synthesis of multiple related proteins. A well-known example is the *lac* operon [3].

Considering the operons, these primary units of regulation of gene expression often comprise additional regulatory DNA sequences. The so-called cis-acting elements, also referred to Transcription Factor Binding Sites (TFBS), which are specific sites where gene transcription regulatory proteins bind to, directly or indirectly affect the initiation of transcription [2]. Also known as regulators or transcription factors (TFs), regulatory proteins are trans-acting elements that either induce or repress the expression of a given gene. A given regulator might coordinate the regulation of many operons. A network of operons with a common regulator is so called a regulon [1].

Besides the fundamental biological machinery described above, there are many other regulatory mechanisms for controlling gene expression in prokaryotes, such as transcriptional attenuation and gene regulation by recombination [1]. Yet, the scope of this review only encompasses biological processes that are quantitatively described in literature, databases and methodological approaches.

2 Regulatory Information Resources

Considering their type, there are two main resources of regulatory information. A considerable number of databases collect transcriptional data regarding elements of biological machinery that control gene expression. Valuable information regarding promoters, TFs, TFBS and operons among other, is often kept in these databases, once data is retrieved from literature or inferred with comparative genomics tools. On the other hand, a restricted number of databases centralize most gene expression data currently available. The development of high-throughput technologies, such as next-generation sequencing, contributed to the increase in the amount of gene expression data found in these public repositories.

Depending on the methodology used for reconstructing integrated genome-scale models of metabolism and gene expression, one may resort to either transcriptional regulatory data, raw gene expression data or a combination of both.

2.1 Databases of Transcriptional Information

Databases of prokaryotic transcriptional information store valuable information on regulatory interactions that take place inside the cell. Information contained in these databases often describes the biological machinery responsible for controlling the gene expression as a function of changes in the environmental conditions.

According to their representativity, databases of regulatory information can be categorized into two groups: organism-specific and non-organism specific. Table S1 of supplementary file 1 (https://bit.ly/2WonbG0) provides the most relevant databases of regulatory information grouped by the corresponding scope.

Comprehensive organism-specific databases are only available for model organisms such as *Escherichia. coli* and *Bacillus subtilis*, or well-known bacteria such as *Mycobacterium tuberculosis*, Gamma-proteobacteria, Mycobacteria and Cyanobacteria. These databases are the result of collaborative task forces aimed at collecting regulatory information on a single organism, which are spread all over literature and other resources, like databases of gene expression data.

For instance, the Database of Transcriptional Regulation in *B. subtilis* (DBTBS) comprises a collection of experimentally validated gene regulatory relations and the corresponding TFBS of the bacterium genes [4]. Recently, the reconstruction of the Transcriptional Regulatory Network (TRN) for *B. subtilis* combined information available in this database with data of less comprehensive databases and a gene expression dataset [5]. This work, by Faria *et al.* [5], also proposed a novel representation of fundamental units of function within a cell called Atomic Regulons (ARs) [6].

Another set of curated regulatory interactions can be obtained for *E. coli* in the RegulonDB [7]. The authors present this database as a unified resource for transcriptional regulation in *E. coli* K-12. In the latest version, an additional effort for incorporating high-throughput-generated binding data was made, extending the understanding of gene expression in the model organism. However, non-model organisms' databases are less comprehensive. Some of these resources comprise regulatory DNA motifs, respective TFs and regulatory networks of less described bacteria.

These databases are gold standards of regulatory information for a single prokaryotic organism. Hence, these should be assessed to infer high-quality TRNs or integrate regulatory information with Genome-Scale Metabolic (GSM) models. Nevertheless, they may lack regulatory interactions found in recent data.

Non-organism specific databases offer limited information. These resources contain information for vast phylogenetic clades, including specific elements of the biological machinery of regulation of gene expression. Comprehensive information, such as the regulatory interactions between TFs and target genes, can, nevertheless, be obtained with comparative genomics approaches.

For example, RegPrecise [8] represents at least 14 taxonomic groups of bacteria, with a collection of transcriptional regulons, determined with comparative genomics approaches, inferred from high-quality manually-curated transcriptional regulatory interactions, namely called regulogs [9]. PRODORIC2 is another database that includes manually curated and unique collection of TFBS for a considerable range of bacteria [10]. Other databases, shown in Table S1, also provide relevant information regarding the regulation of gene expression in prokaryotes, such as putative operons, promoters, TFs and TFBS for multiple species of bacteria. These databases are useful for comparative genomics-based approaches towards the reconstruction of TRN.

2.2 Databases of Gene Expression Data

Up to now, the main sources of gene expression data were based on high-throughput transcriptomics technologies, namely microarrays [11] and RNA-seq [12, 13]. Whereas the former contributed for the initial steps of the research in this area, the latter is responsible for a paradigm shift (see below). Other techniques for measuring gene expression level, such as ChIP-chip [14], SAGE [15], or ChIP-seq [16] are worth mentioning as well.

Expression profiling-based techniques such as microarrays [11] and SAGE [15] allow measuring the level of gene expression and quantifying the amount of mRNA, respectively. Besides the nature of these possible outputs, genome binding experiments also provide insights over DNA-protein binding targets [14]. NGS-based technologies have the advantage of being sensitive while providing whole-genome direct measurements of mRNA without previous knowledge of the genome sequence [12, 13, 16]. These techniques are also able to detect transcription starting sites [17].

Functional genomics data repositories, like GEO [18] and ArrayExpress [19], store gene expression data for a wide diversity of organisms, including bacteria. Furthermore, both databases respect the Minimum Information About a Microarray Experiment (MIAME) [20] and provide query and browsing tools for analyzing and retrieving gene expression data. Other databases of gene expression data derived from microarray and RNA-seq experiments are COLOMBOS [21] and M3D [22]. Both databases provide comprehensive compendia of bacterial gene expression normalized and downstream processed data. These databases are of extreme importance for reconstructing novel TRNs or determining sets of co-expressed genes using *de novo* reverse engineering-based approaches. Besides, the mentioned resources of gene expression data have already been used for reconciling gene expression data with GSM models.

GEO and ArrayExpress were surveyed as these are the major sources of gene expression data to date. The type and amount of available expression studies, as well as availability of NGS-based techniques, are summarized in Fig. 1A and B. The distribution of gene expression bacterial data was also retrieved. As shown in the Figure S1-C and S1-D of the supplementary file 1 (https://bit.ly/2WonbG0), GEO was further analyzed by collecting the availability of experimental series throughout the years and determining the most-represented bacterial species, respectively.

This survey shows that most data available in both databases is from expression profiling and transcription profiling studies, with 82094 (70%) and 8150 (66%) experimental series for GEO and ArrayExpress, respectively (Fig. 1A and B). In 2012, NGS-based studies represented approximately 2% of the data available in GEO [23].

Although most expression series are derived from microarray-based studies, 64% and 77% in GEO and ArrayExpress, respectively, as of February 2019, the proportion of NGS studies (36% and 23%) has risen significantly. These numbers are aligned with predictions for high-throughput sequencing techniques [23]. GEO has seen a consistent increase in publicly available gene expression experimental series since 2012, at a rate of approximately 11000 series a year (Figure S1 – C of the supplementary file 1 available at https://bit.ly/2WonbG0).

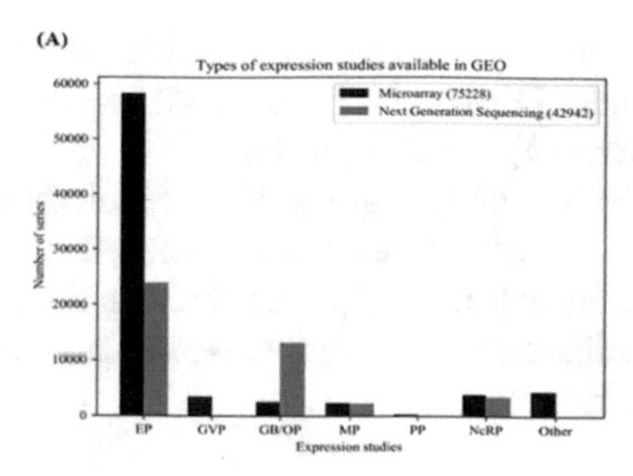

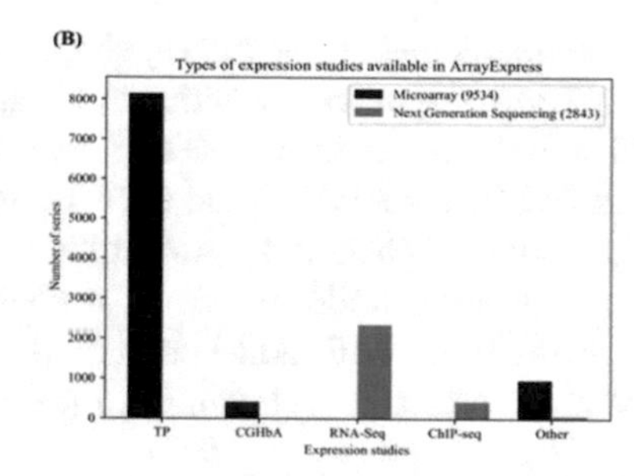

Fig. 1. Survey of the GEO and ArrayExpress databases. Types of available expression studies in GEO (**A**) and ArrayExpress (**B**) for a total of 118170 and 12375 series, respectively. EP (Expression profiling); GVP (Genome Variation Profiling); GB/OP (Genome binding/occupancy profiling); MP (Methylation profiling); PP (Protein profiling); NcRNAP (Non-coding RNA profiling); TP (Transcription profiling); CGHbA (Comparative genomic hybridization by array).

Analyzing the amount of experimental series for each species revealed that the proportion of bacterial-associated data is about 8% and 11% in GEO and ArrayExpress, respectively. As depicted in the Figure S1 – D of the supplementary file 1 (https://bit.ly/2WonbG0), the most-represented bacterial species in GEO as well as in ArrayExpress (data not shown) is *E. coli*. Additionally, we found that Proteobacteria, Firmicutes and Actinobacteria are extensively represented phyla, including 18 of the 20 most-represented bacterial species.

3 Methods for Integrating TRN Reconstruction or Gene Expression Data in GSM Models

As of 2001, several methods have been developed for assisting in the reconstruction and analysis of integrated genome-scale models of metabolism and gene expression [23–25]. The main theory, type of implementation and major drawbacks associated with these methods were addressed, to understand how they comply regulatory information with metabolism. Table S2 of supplementary file 1 (https://bit.ly/2WonbG0) summarizes their main requirements, implementations and drawbacks.

Contrasting with previous reviews [23–25], this review was extended to include more methods. Blazier and Papin [24] reviewed MADE [26], E-FLUX [27] and PROM [28] by highlighting the methods advantages and limitations. Afterwards, Machado and Herrgård [25] revised and evaluated those methods plus tFBA [29] and the method by Lee *et al.* [30], using two gene expression datasets of *E. coli* and one of *Saccharomyces cerevisiae*. Additionally, rFBA [31], SR-FBA [32], PROM [28] and tFBA [29] were already classified and categorized according to the deviations from traditional phenotype simulation with FBA [23]. Besides of these previous reports, TIGER toolbox [33], GIM^{3}E [34], FlexFlux [35], TRFBA [36], CoRegFlux [37] and ME-models [38] were never surveyed before.

Methods were grouped by the type of implementation, namely whether they integrate TRN reconstructions, gene expression data or both. Regardless of classifying these methodologies in toolboxes, simple algorithms, computational tools, advantages or drawbacks, the main deviations to the standard constraint-based modeling approach and Flux Balance Analysis (FBA) framework [39] are also presented.

Unlike ME-models [38] and GIM^{3}E [34], all methods addressed in this study allow the simulation of phenotypes for multiple environmental and genetic conditions using integrated genome-scale models of metabolism and transcriptional information. Note that, GIMM^{3}E and ME-models require the utilization of additional omics data such as exometabolomics and proteomics, respectively.

4 Discussion

The present study is aimed at highlighting the required resources, features and limitations of the latest efforts towards the reconstruction of integrated genome-scale models of metabolism and gene expression. The existence of new sources of regulatory information and gene expression data (e.g. RNA-seq and ChIP-seq) opens previously closed doors for introducing new methods. Although the main share of gene expression data is still from microarray expression studies, the number of datasets and series obtained by NGS-based technologies is on the rise. Nevertheless, most of the methods surveyed in this article, except the method by Lee *et al.* [30], only used microarray expression data.

None of the methods for integrating gene expression with metabolic models, previously evaluated by Machado and coworkers [25], outperforms each other in phenotype predictions. Furthermore, simple growth maximization with parsimony FBA (pFBA) [40], performed as well as the evaluated methods, namely MADE [26], E-FLUX [27], PROM [28], tFBA [29] and the method by Lee *et al.* [30]. This indicates that the promising results reported by these methods are just mere artifacts.

In fact, most results presented by such tools might be related with rigid constraints created around the nature of the gene expression dataset. Whereas some methods require large-scale gene expression datasets to be robust, others resort to mapping levels of gene expression directly with the reactions bounds which might not be the case for all organisms and datasets. The first methods to ever be developed (RFBA [31] and SR-FBA [32]) limit the solution space by removing possible solutions with Boolean logic. Complex formulations, requirements for large-scale or specific gene expression datasets (that are scarce for some bacteria groups) and incongruences obtained in recent benchmarking tests, pose a hard challenge for using these methods out of the scope they were developed for.

Some of the major drawbacks are not a repercussion of the methodologies itself but rather due to the difficulty in propagating their implementations to other organisms, especially those poorly documented. Hence, the perspective of reconstructing integrated genome-scale models of metabolism and gene expression for diverse prokaryotes rather than well-known organisms is still a complex endeavor. Nevertheless, this gap can be overcome by novel approaches, such as using comparative genomics tools for determining *in silico* regulatory mechanisms that affect metabolism [23].

A user-friendly tool implementing different methods and approaches as a function of the available data would likely shed some light on the reconstruction of these methods. Lastly, reconstructing models that incorporate TRN reconstructions should be more advantageous when compared with only the reconciliation of gene expression data into a new FBA-based formulation. Firstly, the model would provide comprehensive knowledge regarding the metabolic and regulatory events occurring inside the cell. Secondly, the various approaches as well as the amount of data available for reconstructing and integrating TRN into GSM models would ease the diffusion of this approach to most bacteria having a sequenced genome. This hypothetical computational tool would therefore be able to combine different sources of regulatory information available in the resources discussed above, which are rarely combined.

Acknowledgements. This study was supported by the Portuguese Foundation for Science and Technology (FCT) under the scope of the strategic funding of UID/BIO/04469/2019 unit and BioTecNorte operation (NORTE-01-0145-FEDER-000004) funded by the European Regional Development Fund under the scope of Norte2020-Programa Operacional Regional do Norte. Fernando Cruz holds a doctoral fellowship (SFRH/BD/139198/2018) funded by the FCT. The authors thank project SHIKIFACTORY100 - Modular cell factories for the production of 100 compounds from the shikimate pathway (814408) funded by the European Commission.

References

1. Nelson, D.L., Cox, M.M.: Lehninger Principles of Biochemistry. W.H. Freeman, New York (2008)
2. Lodish, H., Berk, A., Zipursky, S.L., Matsudaira, P., Baltimore, D., Darnell, J.: Molecular Cell Biology. W.H. Freeman, New York (2000)
3. Willson, C., Pebrin, D., Cohn, M., Jacob, F., Monod, J.: Non-inducible mutants of the regulator gene in the "lactose" system of Escherichia coli. J. Mol. Biol. **8**, 582–592 (1964)
4. Sierro, N., Makita, Y., de Hoon, M., Nakai, K.: DBTBS: a database of transcriptional regulation in Bacillus subtilis containing upstream intergenic conservation information. Nucleic Acids Res. **36**, D93–D96 (2008)
5. Faria, J.P., Overbeek, R., Taylor, R.C., Conrad, N., Vonstein, V., Goelzer, A., Fromion, V., Rocha, M., Rocha, I., Henry, C.S.: Reconstruction of the regulatory network for bacillus subtilis and reconciliation with gene expression data. Front. Microbiol. **7**, 275 (2016)
6. Faria, J.P., Davis, J.J., Edirisinghe, J.N., Taylor, R.C., Weisenhorn, P., Olson, R.D., Stevens, R.L., Rocha, M., Rocha, I., Best, A.A., DeJongh, M., Tintle, N.L., Parrello, B., Overbeek, R., Henry, C.S.: Computing and applying atomic regulons to understand gene expression and regulation. Front. Microbiol. **7**, 1819 (2016)
7. Santos-Zavaleta, A., Sánchez-Pérez, M., Salgado, H., Velázquez-Ramírez, D.A., Gama-Castro, S., Tierrafría, V.H., Busby, S.J.W., Aquino, P., Fang, X., Palsson, B.O., Galagan, J.E., Collado-Vides, J.: A unified resource for transcriptional regulation in Escherichia coli K-12 incorporating high-throughput-generated binding data into RegulonDB version 10.0. BMC Biol. **16**, 91 (2018)
8. Novichkov, P.S., Brettin, T.S., Novichkova, E.S., Dehal, P.S., Arkin, A.P., Dubchak, I., Rodionov, D.A.: RegPrecise web services interface: programmatic access to the transcriptional regulatory interactions in bacteria reconstructed by comparative genomics. Nucleic Acids Res. **40**, W604–W608 (2012)

9. Novichkov, P.S., Kazakov, A.E., Ravcheev, D.A., Leyn, S.A., Kovaleva, G.Y., Sutormin, R. A., Kazanov, M.D., Riehl, W., Arkin, A.P., Dubchak, I., Rodionov, D.A.: RegPrecise 3.0 – a resource for genome-scale exploration of transcriptional regulation in bacteria. BMC Genom. **14**, 745 (2013)
10. Eckweiler, D., Dudek, C.-A., Hartlich, J., Brötje, D., Jahn, D.: PRODORIC2: the bacterial gene regulation database in 2018. Nucleic Acids Res. **46**, D320–D326 (2018)
11. Young, R.A.: Biomedical discovery with DNA arrays. Cell **102**, 9–15 (2000)
12. Mortazavi, A., Williams, B.A., McCue, K., Schaeffer, L., Wold, B.: Mapping and quantifying mammalian transcriptomes by RNA-Seq. Nat. Methods **5**, 621–628 (2008)
13. Nagalakshmi, U., Wang, Z., Waern, K., Shou, C., Raha, D., Gerstein, M., Snyder, M.: The transcriptional landscape of the yeast genome defined by RNA sequencing. Science (80-.) **320**, 1344–1349 (2008)
14. Iyer, V.R., Horak, C.E., Scafe, C.S., Botstein, D., Snyder, M., Brown, P.O.: Genomic binding sites of the yeast cell-cycle transcription factors SBF and MBF. Nature **409**, 533–538 (2001)
15. Velculescu, V.E., Zhang, L., Zhou, W., Vogelstein, J., Basrai, M.A., Bassett, D.E., Hieter, P., Vogelstein, B., Kinzler, K.W.: Characterization of the yeast transcriptome. Cell **88**, 243–251 (1997)
16. Johnson, D.S., Mortazavi, A., Myers, R.M., Wold, B.: Genome-wide mapping of in vivo protein-DNA interactions. Science (80-.) **316**, 1497–1502 (2007)
17. Price, M.N., Deutschbauer, A.M., Kuehl, J.V., Liu, H., Witkowska, H.E., Arkin, A.P.: Evidence-based annotation of transcripts and proteins in the sulfate-reducing bacterium Desulfovibrio vulgaris Hildenborough. J. Bacteriol. **193**, 5716–5727 (2011)
18. Barrett, T., Wilhite, S.E., Ledoux, P., Evangelista, C., Kim, I.F., Tomashevsky, M., Marshall, K.A., Phillippy, K.H., Sherman, P.M., Holko, M., Yefanov, A., Lee, H., Zhang, N., Robertson, C.L., Serova, N., Davis, S., Soboleva, A.: NCBI GEO: archive for functional genomics data sets—update. Nucleic Acids Res. **41**, D991–D995 (2012)
19. Kolesnikov, N., Hastings, E., Keays, M., Melnichuk, O., Tang, Y.A., Williams, E., Dylag, M., Kurbatova, N., Brandizi, M., Burdett, T., Megy, K., Pilicheva, E., Rustici, G., Tikhonov, A., Parkinson, H., Petryszak, R., Sarkans, U., Brazma, A.: ArrayExpress update—simplifying data submissions. Nucleic Acids Res. **43**, D1113–D1116 (2015)
20. Brazma, A., Hingamp, P., Quackenbush, J., Sherlock, G., Spellman, P., Stoeckert, C., Aach, J., Ansorge, W., Ball, C.A., Causton, H.C., Gaasterland, T., Glenisson, P., Holstege, F.C.P., Kim, I.F., Markowitz, V., Matese, J.C., Parkinson, H., Robinson, A., Sarkans, U., Schulze-Kremer, S., Stewart, J., Taylor, R., Vilo, J., Vingron, M.: Minimum information about a microarray experiment (MIAME)—toward standards for microarray data. Nat. Genet. **29**, 365–371 (2001)
21. Moretto, M., Sonego, P., Dierckxsens, N., Brilli, M., Bianco, L., Ledezma-Tejeida, D., Gama-Castro, S., Galardini, M., Romualdi, C., Laukens, K., Collado-Vides, J., Meysman, P., Engelen, K.: COLOMBOS v3.0: leveraging gene expression compendia for cross-species analyses. Nucleic Acids Res. **44**, D620–D623 (2016)
22. Faith, J.J., Driscoll, M.E., Fusaro, V.A., Cosgrove, E.J., Hayete, B., Juhn, F.S., Schneider, S. J., Gardner, T.S.: Many Microbe Microarrays Database: uniformly normalized Affymetrix compendia with structured experimental metadata. Nucleic Acids Res. **36**, D866–D870 (2007)
23. Faria, J.P., Overbeek, R., Xia, F., Rocha, M., Rocha, I., Henry, C.S.: Genome-scale bacterial transcriptional regulatory networks: reconstruction and integrated analysis with metabolic models. Brief. Bioinform. **15**, 592–611 (2014)
24. Blazier, A.S., Papin, J.A.: Integration of expression data in genome-scale metabolic network reconstructions. Front. Physiol. **3**, 299 (2012)

25. Machado, D., Herrgård, M.: Systematic evaluation of methods for integration of transcriptomic data into constraint-based models of metabolism. PLoS Comput. Biol. **10**, e1003580 (2014)
26. Jensen, P.A., Papin, J.A.: Functional integration of a metabolic network model and expression data without arbitrary thresholding. Bioinformatics **27**, 541–547 (2011)
27. Colijn, C., Brandes, A., Zucker, J., Lun, D.S., Weiner, B., Farhat, M.R., Cheng, T.-Y., Moody, D.B., Murray, M., Galagan, J.E.: Interpreting expression data with metabolic flux models: predicting mycobacterium tuberculosis mycolic acid production. PLoS Comput. Biol. **5**, e1000489 (2009)
28. Chandrasekaran, S., Price, N.D.: Probabilistic integrative modeling of genome-scale metabolic and regulatory networks in Escherichia coli and Mycobacterium tuberculosis. Proc. Natl. Acad. Sci. U.S.A **107**, 17845–17850 (2010)
29. van Berlo, R.J.P., de Ridder, D., Daran, J.-M., Daran-Lapujade, P.A.S., Teusink, B., Reinders, M.J.T.: Predicting metabolic fluxes using gene expression differences as constraints. IEEE/ACM Trans. Comput. Biol. Bioinform. **8**, 206–216 (2011)
30. Lee, D., Smallbone, K., Dunn, W.B., Murabito, E., Winder, C.L., Kell, D.B., Mendes, P., Swainston, N.: Improving metabolic flux predictions using absolute gene expression data. BMC Syst. Biol. **6**, 73 (2012)
31. Covert, M.W., Schilling, C.H., Palsson, B.: Regulation of gene expression in flux balance models of metabolism. J. Theor. Biol. **213**, 73–88 (2001)
32. Shlomi, T., Eisenberg, Y., Sharan, R., Ruppin, E.: A genome-scale computational study of the interplay between transcriptional regulation and metabolism. Mol. Syst. Biol. **3**, 101 (2007)
33. Jensen, P.A., Lutz, K.A., Papin, J.A.: TIGER: Toolbox for integrating genome-scale metabolic models, expression data, and transcriptional regulatory networks. BMC Syst. Biol. **5**, 147 (2011)
34. Schmidt, B.J., Ebrahim, A., Metz, T.O., Adkins, J.N., Palsson, B.Ø., Hyduke, D.R.: GIM3E: condition-specific models of cellular metabolism developed from metabolomics and expression data. Bioinformatics **29**, 2900–2908 (2013)
35. Marmiesse, L., Peyraud, R., Cottret, L.: FlexFlux: combining metabolic flux and regulatory network analyses. BMC Syst. Biol. **9**, 93 (2015)
36. Motamedian, E., Mohammadi, M., Shojaosadati, S.A., Heydari, M.: TRFBA: an algorithm to integrate genome-scale metabolic and transcriptional regulatory networks with incorporation of expression data. Bioinformatics. **33**, btw772 (2017)
37. Banos, D.T., Trébulle, P., Elati, M.: Integrating transcriptional activity in genome-scale models of metabolism. BMC Syst. Biol. **11**, 134 (2017)
38. Lloyd, C.J., Ebrahim, A., Yang, L., King, Z.A., Catoiu, E., O'Brien, E.J., Liu, J.K., Palsson, B.O.: COBRAme: a computational framework for genome-scale models of metabolism and gene expression. PLoS Comput. Biol. **14**, e1006302 (2018)
39. Orth, J.D., Thiele, I., Palsson, B.O.: What is flux balance analysis? Nat. Publ. Gr. **28**, 245–248 (2010)
40. Lewis, N.E., Hixson, K.K., Conrad, T.M., Lerman, J.A., Charusanti, P., Polpitiya, A.D., Adkins, J.N., Schramm, G., Purvine, S.O., Lopez-Ferrer, D., Weitz, K.K., Eils, R., König, R., Smith, R.D., Palsson, B.Ø.: Omic data from evolved E. coli are consistent with computed optimal growth from genome-scale models. Mol. Syst. Biol. **6**, 390 (2010)

Author Index

F. Fdez-Riverola et al. (Eds.): PACBB 2019, AISC 1005, pp. 183–184, 2020.
https://doi.org/10.1007/978-3-030-23873-5

Printed in the United States
By Bookmasters